Level 5

5级

美国原版青少年核心能力拓展

好玩的数学

Targeted Mathematics Student Guided Practice Book

主 编：〔美〕莎拉·约翰逊

天津出版传媒集团
天津科技翻译出版有限公司

著作权合同登记号:图字:02-2013-308

图书在版编目(CIP)数据

美国原版青少年核心能力拓展：好玩的数学=Targeted mathematics: student guided practice book. 5级:英文/（美)约翰逊(Johnson, S.)主编. —天津:天津科技翻译出版有限公司, 2014.4

ISBN 978-7-5433-3346-8

Ⅰ.①美… Ⅱ.①约… Ⅲ.①数学—少儿读物 ②英语—口语—少儿读物 Ⅳ.①O1-49 ②H319.9-49

中国版本图书馆 CIP 数据核字(2013)第 320401 号

授权单位: Teacher Created Materials, Inc.
出　　版: 天津科技翻译出版有限公司
出 版 人: 刘 庆
地　　址: 天津市南开区白堤路 244 号
邮政编码: 300192
电　　话: 022-87894896
传　　真: 022-87893237
网　　址: www. tsttpc. com
印　　刷: 唐山天意印刷有限责任公司
发　　行: 全国新华书店
版本记录: 880×1230　16 开本　10.5 印张　210 千字
2014 年 4 月第 1 版　2014 年 4 月第 1 次印刷
定价:25.00 元

(如发现印装问题,可与出版社调换)

编者前言

《美国原版青少年核心能力拓展：好玩的数学》(学前阶段及1~8级)是从美国教师创新教材出版公司(Teacher Created Materials Inc.)引进的现行介入式数学指南，全面反映了美国学前班至八年级数学课的现行教学内容及教学要求，同时也为我们展示了美国青少年丰富多彩、生动活泼的学习场景。

阅读使用这套丛书会让你有一种犹如在美国上学、与美国的小朋友同步学习的亲身体验。从中不仅可以了解美国学生在数学课上学些什么，做些什么样的作业，考些什么样的数学题；还可以知道他们的老师在课堂上讲些什么，以及对学生的要求是什么。由此你会发现，他们的数学课与我们的有相同之处，但也并非完全相同。我们侧重于背公式，做习题，备考应试；而他们侧重于理解和掌握数学的基础知识，既讲述初等数学的内容，又介绍了一些高等数学、数论、概率论、统计学的知识，并与其他学科相互联系，从而了解数学在其他学科中的应用，而且在教学中注意联系实际，注重实践应用，因此上数学课不会让学生感到枯燥乏味，而是感觉生动有趣。二者有着不同的教学理念和方式，如果能通过这套丛书的学习将二者有机地结合起来，取长补短，优势互补，必能开阔你的眼界，提高你对数学概念的理解，提升你的应用能力(当然也包括应试能力)。

数学是世界各地通用的一门学科，有着共同的概念、公式、术语、习题、计算方法，因此在这套书中有着非常熟悉的学习内容和知识背景：学过的数学知识，做过的数学习题，考过的数学试题。特别之处在于这套丛书以英文原版形式体现，这就为你营造了一个在熟悉的背景下学习英语的环境，学会用地道的英语来表达学过的知识，表达真实的日常生活和学习活动，学会用英语和同学进行学习互动，从而大幅度提高你的英语水平。既学了数学又学了英语(而且是非常实用的英语)，岂非两全其美的好事。

这套丛书适用于我国广大青少年读者，尤其是双语学校的学生以及打算到英语国家上高中、上大学的学生。学习这套丛书，就等于在国内体验了国外的学校生活，这对今后的深造无疑是大有裨益的。

打开书本，开启你在国内“留学”的全新生活吧！

英语就得天天练——阅读美国孩子的课余英文原版书

好玩的数学——体验美国青少年数学学习的乐趣

Table of Contents

目 录

Table of Contents

目 录

Student Welcome Letter
欢 迎 信

Dear Student,

You are starting a math program that will help you review fifth grade. Up to this point in school, you have learned many mathematical skills. This program will help you focus on what you already know how to do in math and what you need to learn. You will learn the important mathematical concepts, skills, and vocabulary so that you are ready for sixth grade.

Sometimes students have trouble learning math. It can seem confusing. This program will help you practice math every day. You will review many things you learned in fourth grade. Some of those things include probability, fractions, and problem solving.

Please sign the bottom of this letter and keep it as a reminder of the skills you learned in this math program. Have fun!

____

Student Signature（学生签名）

Diagnostic Test
摸底测试

1 What is the written expression（书面表达）for 16,342 + 127?

Ⓐ sixteen thousand, four hundred sixty-nine

Ⓑ sixteen thousand, four hundred seventy-nine

Ⓒ fifteen thousand, five hundred sixty-nine

Ⓓ fourteen thousand, four hundred sixty-nine

2 Which of these answers shows how to break down（分解）22 x 6 into math facts（数学式）you know?

Ⓕ (10 x 6) – (10 x 2)

Ⓖ (6 x 10) + (6 x 10)

Ⓗ (6 x 20) + (6 x 2)

Ⓙ (11 x 6) + (12 x 6)

3 Use any division method to find the quotient（商）.

$8\overline{)967}$

Ⓐ 121 R7（余7）

Ⓑ 120 R7

Ⓒ 140 R7

Ⓓ 121 R6

4 Irma bought a pair of tennis shoes（网球鞋）for $39.65 and a pair of boots（靴子）for $51.39. How much did she spend in all?

Ⓕ $91.04

Ⓖ $101.05

Ⓗ $92.05

Ⓙ $89.04

5 Which of the following has the least value?

Ⓐ 0.243

Ⓑ 0.21

Ⓒ 0.198

Ⓓ 0.3

6 Choose the correct answer.

$0.80 x 1,000 = ____________

Ⓕ $8.00

Ⓖ $800

Ⓗ $80.00

Ⓙ $8,000

Diagnostic Test *(cont.)*

7 Which of these answers lists all of the factors of 8?

Ⓐ 2, 4, 8

Ⓑ 2, 4

Ⓒ 2, 4, 6, 8

Ⓓ 1, 2, 4, 8

10 Simplify（约分） the fraction below:

$\frac{18}{81}$

Ⓕ $\frac{9}{40}$

Ⓖ $\frac{1}{4}$

Ⓗ $\frac{3}{13}$

Ⓙ $\frac{2}{9}$

8 Alex added two prime numbers（质数）. The sum（和） was an odd number（奇数）. Which of these numbers had to be one of the prime numbers?

Ⓕ 2

Ⓖ 3

Ⓗ 5

Ⓙ 7

11 Kevin walked $\frac{25}{8}$ miles（英里）. Simplify the improper fraction（可约分数）.

Ⓐ $\frac{12}{4}$ miles

Ⓑ 3 miles

Ⓒ $2\frac{10}{8}$ miles

Ⓓ $3\frac{1}{8}$ miles

*为体现原版书的特色，书中出现的计量单位在不影响解题思路的情况下均保持原貌，个别在我国不常用的计量单位已做换算标注。计量单位换算可参见附录B：数学用表。

9 Nihal and Monique went to Pizza Place. Nihal ate $\frac{4}{8}$ of the pizza. Monique ate $\frac{2}{8}$ of the pizza. How much pizza did they eat?

Ⓐ $\frac{2}{8}$

Ⓑ $\frac{1}{8}$

Ⓒ $\frac{6}{8}$

Ⓓ $\frac{7}{8}$

12 Determine the next two numbers in this sequence.

4, 9, 15, 22, 30, ______, ______

Ⓕ 35, 41

Ⓖ 38, 46

Ⓗ 37, 45

Ⓙ 39, 49

Diagnostic Test *(cont.)*

13 The length of a poster（海报） is 3 feet. The area is 21 feet squared（平方英尺）. Which equation best represents the width（宽） of the poster?

Ⓐ $21 \cdot w = 3$

Ⓑ $3 \cdot w = 21$

Ⓒ $w \div 3 = 21$

Ⓓ $21 \cdot 3 = w$

14 Marco and his family drove 340 miles on their vacation（假期）. They stopped after 150 miles to see the Grand Canyon（科罗拉多大峡谷）. Which equation best represents the number of miles they traveled after that stop?

Ⓕ $150 \times m = 340$

Ⓖ $340 \div m = 150$

Ⓗ $m - 150 = 340$

Ⓙ $150 + m = 340$

15 What is the perimeter（周长） of a figure with these 4 sides: 3 cm, 5 cm, 7 cm, 4 cm?

Ⓐ 23 cm

Ⓑ 15 cm

Ⓒ 19 cm

Ⓓ 16 cm

16 What is the best estimate for the length of a crayon（蜡笔）?

Ⓕ 3 inches

Ⓖ 3 centimeters

Ⓗ 3 millimeters

Ⓙ 3 feet

*本题单位换算参见附录B：数学用表。

17 What is the perimeter of a square eraser（橡皮） that is 5.6 cm long on each side?

Ⓐ 22.2 cm

Ⓑ 20.24 cm

Ⓒ 22.4 cm

Ⓓ 11.2 cm

18 Compute the area of a desk that is 12 ft. long and 13 ft. wide.

Ⓕ 154 $ft.^2$

Ⓖ 25 $ft.^2$

Ⓗ 50 $ft.^2$

Ⓙ 156 $ft.^2$

Diagnostic Test *(cont.)*

19 What is the volume (体积) of the shape?

- Ⓐ 18 units squared (正方形)
- Ⓑ 16 units cubed (正方体)
- Ⓒ 18 units cubed
- Ⓓ 16 units squared

20 What are the coordinates (坐标) for point B?

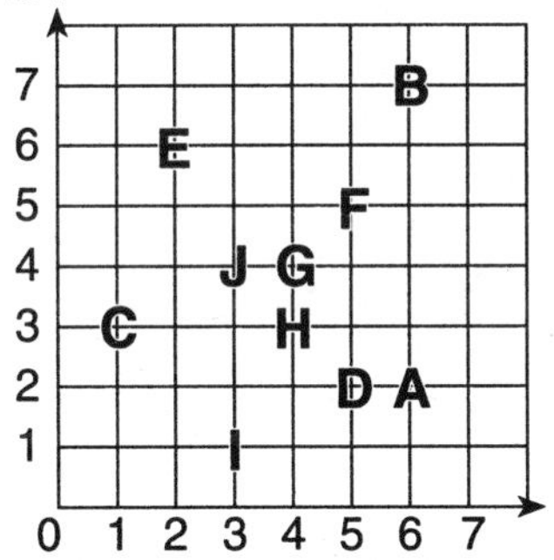

- Ⓕ (7, 6)
- Ⓖ (6, 7)
- Ⓗ (6, 6)
- Ⓙ (7, 5)

21 Which two lines are parallel (平行)?

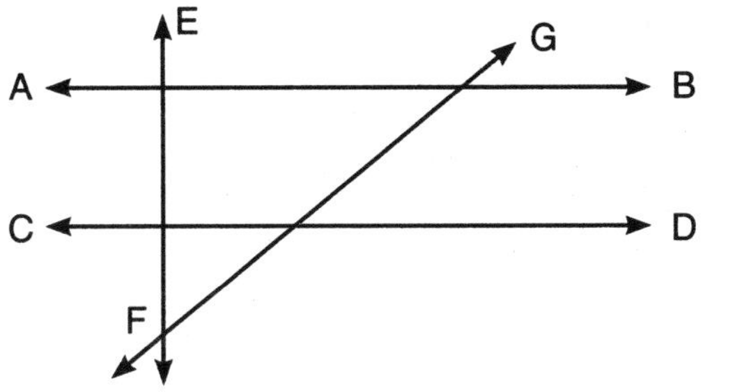

- Ⓐ $\overleftrightarrow{AB}$ and $\overleftrightarrow{EF}$
- Ⓑ $\overleftrightarrow{EF}$ and $\overleftrightarrow{CD}$
- Ⓒ $\overleftrightarrow{AB}$ and $\overleftrightarrow{CD}$
- Ⓓ None

22 These shapes are ______________ .

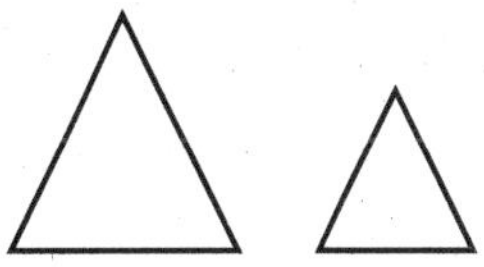

- Ⓕ right triangle (直角三角形)
- Ⓖ congruent (全等三角形)
- Ⓗ similar (相似三角形)
- Ⓙ quadrilaterals (四边形)

23 This shows an example of a _____ .

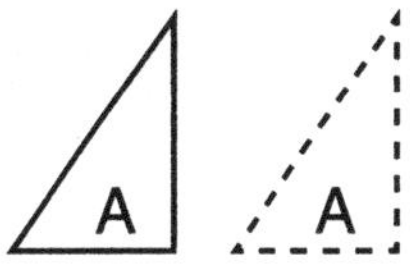

- Ⓐ reflection (反射)
- Ⓑ flip (翻转)
- Ⓒ rotation (旋转)
- Ⓓ translation (平移)

24 Which shape fits in (适应) area C of the Venn diagram (维恩图)?

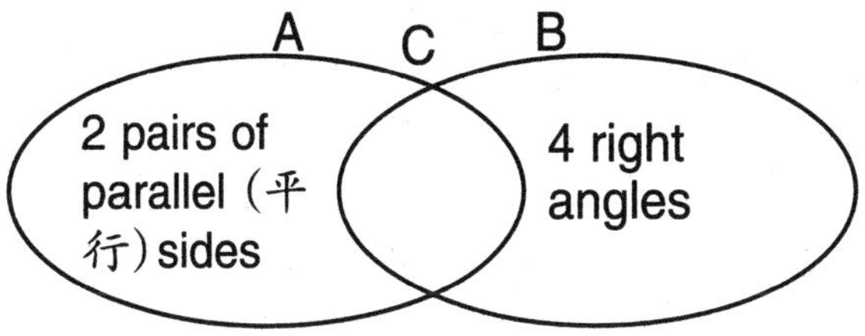

- Ⓕ rectangle (矩形)
- Ⓗ triangle (三角形)
- Ⓖ rhombus (菱形)
- Ⓙ trapezoid (梯形)

Diagnostic Test *(cont.)*

25 What type of triangle is this?

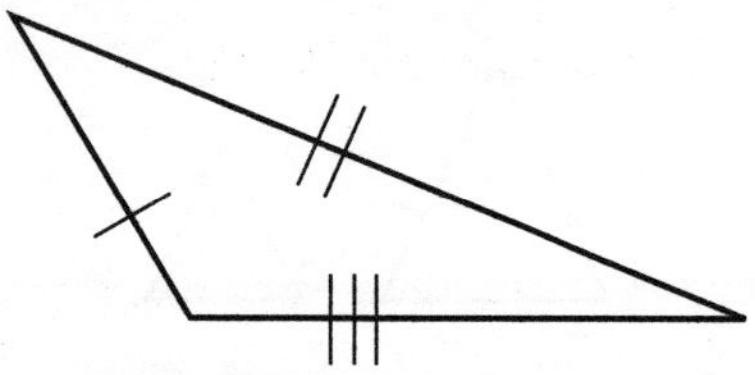

(A) equilateral（等边）

(B) acute（锐角）

(C) right（直角）

(D) scalene（不等边）

26 Mike recorded these heights in inches for 10 of his classmates: 60, 45, 52, 54, 53, 60, 57, 56, 57, 60. Which is the mode（众数）?

(F) 60

(G) 57

(H) 56

(J) 45

27 How many pizzas did Jack's sell on Friday?

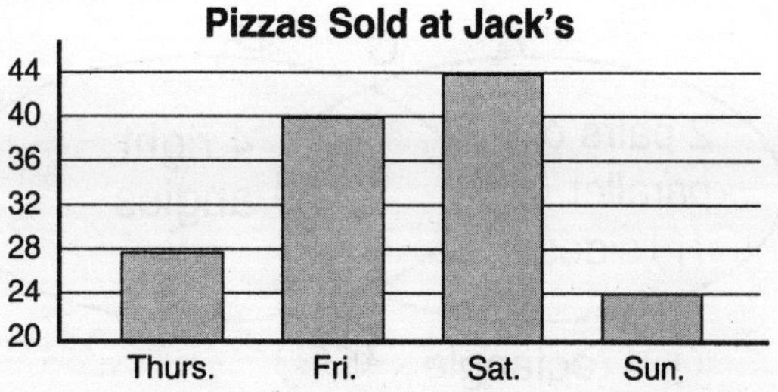

(A) 28 pizzas　　(C) 24 pizzas

(B) 40 pizzas　　(D) 44 pizzas

28 What was the average（平均）monthly temperature for March shown on the line graph?

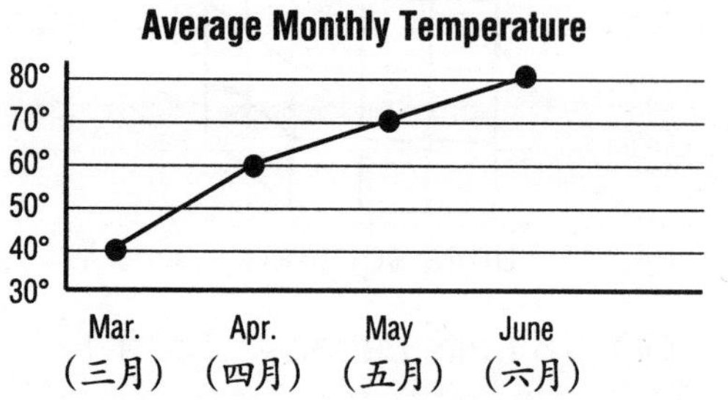

(F) 30°　　(H) 40°

(G) 35°　　(J) 45°

29 There are more soccer balls（足球）than ______.

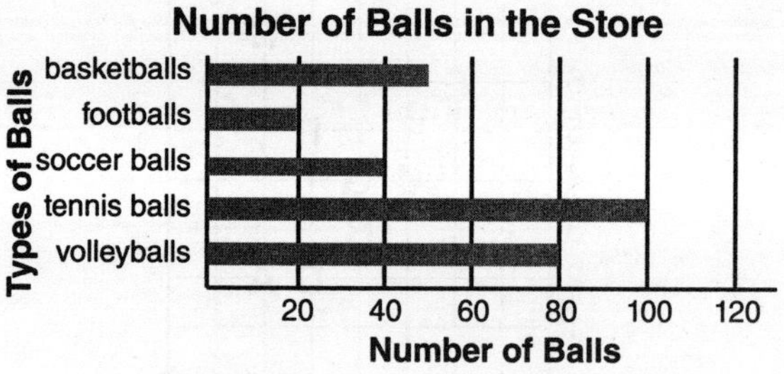

(A) footballs（橄榄球）　　(C) basketballs（篮球）

(B) volleyballs（排球）　　(D) tennis balls（网球）

30 Choose the best category（范畴）for this statement: *You will breathe air in the next five minutes.*

(F) Impossible（不可能）

(G) Unlikely（不太可能）

(H) Likely（很可能）

(J) Certain（必然）

Place Value Grid
位值表格

Directions: Write numbers in the grid（格子）below. Place the digits（数字）in the correct columns（列）and then identify what each digit represents. For example: 2,653 = 2 thousands, 6 hundreds, 5 tens, and 3 ones.

	Ten Thousands（万）	Thousands（千）	Hundreds（百）	Tens（十）	Ones（一）
1.					
2.					
3.					
4.					
5.					
6.					

Directions: On the lines below, write each of the above numbers in words. (For example: 2,653 = two thousand, six hundred fifty-three.)

1. ______________________________

____

2. ______________________________

____

3. ______________________________

____

4. ______________________________

____

5. ______________________________

____

6. ______________________________

____

Guessing and Checking
估算与验算

Guessing and checking can be used to solve a variety of problems. Begin with an educated（有根据的）guess, a guess based on information you already know, or ideas that make sense. Don't make a wild guess. Always base your first guess on important facts.

Check your guess against the facts and information in the problem. If your guess is wrong, change it according to whether it is too small or too large. Check your answer again. Repeat these steps until you find the solution.

Learning how to make reasonable guesses takes practice. Creating a table will help you keep track of your guesses and results until you find the correct solution.

Noting the Important Facts

Begin by looking for the important facts in the problem and determining what you need to solve.

Problem: Sticker Fun!（贴纸游戏）

The Problem

Mr. Collins handed out（分发）45 reward（奖励）stickers to his class over a five-day period. Each day, he handed out three more stickers than the day before.

How many stickers did Mr. Collins give to his students each day?

Understanding the Problem

- *What do we know?*

  The total number of stickers is 45.

  The stickers were handed out over a five-day period. Each day, more stickers are handed out.

- *What do we need to find out?*

  How many stickers did Mr. Collins hand out each day?

Planning and Communicating a Solution（计划和沟通解决方案）

To make an educated guess, you need a starting point（起始点）. For example, it might be useful to start with five stickers. In your table, keep track of the total stickers so that you know when the total of 45 stickers is reached. Create a table with three rows and six columns.

Guess 1

Day	1	2	3	4	5
Number of Stickers	5	8	11	14	17
Total	5	13	24	38	55

This first guess is too high because there are 55 total after 5 days. Start with a lower number of stickers for the next guess.

Guess 2

Day	1	2	3	4	5
Number of Stickers	2	5	8	11	14
Total	2	7	15	26	40

The total is too low. For guess 3, start with a number between 2 and 5.

Guess 3

Day	1	2	3	4	5
Number of Stickers	3	6	9	12	15
Total	3	9	18	30	45

- *Do you see the answer?*

  Guess 3 is correct. Mr. Collins handed out 3 stickers the first day, 6 on the second day, 9 on the third day, 12 on the fourth day, and 15 on the fifth day.

Reflecting and Generalizing（指导和概括）

Could the problem have been solved any other way?

Extension

If Mr. Collins continued this pattern the next week, how many stickers did he give out by the ninth day?

Standardized Test Preparation 1

标准考前测试 1

1 Which number means eight ten thousands, seven thousands, five hundreds, six tens, and nine ones?

Ⓐ 8,756

Ⓑ 87,659

Ⓒ 875,609

Ⓓ 87,569

2 Which of these answers shows how to break down（分解） 8 x 19 into math facts you know?

Ⓕ (8 x 10) – (8 x 9)

Ⓖ (8 x 12) + (8 x 9)

Ⓗ (8 x 10) + (8 x 8)

Ⓙ (8 x 10) + (8 x 9)

3 Compute to find the quotient（商）.

287 ÷ 9 = ___

Ⓐ 33 R7

Ⓑ 31 R7

Ⓒ 32 R1

Ⓓ 31 R8

4 Christina rode the Whirly Bird Twister, which cost \$3.75, and the Screamin' Coaster, which cost \$2.65. How much did she spend?

Ⓕ \$6.20

Ⓖ \$6.40

Ⓗ \$6.85

Ⓙ \$5.40

5 Which of the decimals（小数） listed has the greatest value?

Ⓐ 0.55

Ⓑ 0.49

Ⓒ 0.2

Ⓓ 0.381

6 Pick one question from this test. Explain how and why you chose your answer.

____

____

____

____

____

____

Use It!

分解运算

Directions: Solve the multiplication（乘法）problems below by breaking them down. Show your work. The first one has been done for you.

1. 15 x 4 = (10 x 4) + (5 x 4) = 40 + 20 = 60
2. 17 x 3 = ______
3. 14 x 5 = ______
4. 25 x 3 = ______
5. 26 x 20 = ______
6. 48 x 12 = ______
7. 200 x 20 = ______
8. 600 x 30 = ______
9. 21 x 6 = ______
10. 19 x 7 = ______
11. 25 x 30 = ______
12. 31 x 30 = ______
13. 48 x 20 = ______
14. 300 x 40 = ______
15. 800 x 50 = ______
16. 1,200 x 70 = ______
17. 2,400 x 60 = ______
18. 3,500 x 80 = ______

School Uniforms

校 服

Most schools have dress codes（着装要求）for students. And now, many schools are requiring parents to purchase uniforms（制服）for their children to wear to school.

Of course, wearing a school uniform is nothing new. Many private schools（私立学校）have had uniforms for years. Private school uniforms are usually ordered from specialty catalogs. Since the mid-1990s, some public schools have started to require students to wear uniforms.

Most school uniforms consist of white shirts with short or long sleeves（袖子）and dark pants（裤子）, shorts（短裤）, or skirts（裙子）. Girls may also be allowed to wear dark jumpers（连衣裤）. Both boys and girls may also wear dark sweaters（毛衣）. The uniforms public schools are requiring often do not have to be specially ordered and can be purchased easily at discount stores（打折店）and department stores（百货商店）.

School districts（学区）have argued（辩论）that uniforms have several positive effects（积极作用）. Requiring students to wear uniforms reduces student competition（竞争）over clothes. Since students are wearing similar（相似）clothes, they can concentrate（注意）more on their schoolwork（家庭作业）rather than what they or their friends are wearing. School officials say the uniforms are more cost efficient, too. School uniforms tend to cost less than most trendy（时髦）clothes. Also, when students are required to wear uniforms, they do not have to buy new clothes to keep up with current fads（时尚）. The uniforms can be passed down to younger siblings（学弟学妹）once a student grows out of them.

Some groups are opposed to school uniforms. They agree that students should have to adhere（遵守, 坚持）to a dress code, but they argue that students should be free to choose the types of clothes they want to wear.

School Uniforms *(cont.)*

What Is the Problem?

It is back-to-school (返校) time, and your school is requiring students to wear uniforms. Your job is to find the costs of the various pieces of the school uniform that the students at your school are going to wear.

You will be asked to work on the problems below, using the guessing and checking strategy. Use the information below to answer the questions.

Before you begin solving each problem, be sure to locate the information you will need. Think about what information you can use to help guide your guesses. Then, you may want to organize your guesses into a table to keep track of them.

Remember: If you have trouble planning ways to solve each problem, reread it and rethink your strategy. When you have a solution, think about whether your answer makes sense.

Problem-Solving Strategy: Guessing and Checking

Problem A

A total of $16 was spent on a skirt and sweater. The sweater cost $4 more than the skirt. How much did each item of clothing cost?

Problem B

A total of $15 was spent on a shirt with long sleeves and a pair of pants. The pair of pants cost $4 more than the shirt. How much did each cost?

A pair of pants and a sweater cost $19.50. The pants cost 50 cents less than the sweater. How much did each cost?

Hint: The cost of each item might not be a whole number.

Problem C

Think about what items of clothing would be needed for a whole week. Use the calculated costs from Problems A and B. Consider that the short-sleeved shirt costs $2 less than a long-sleeved shirt. A pair of shorts costs the same as the skirt in Problem A.

Guess how many outfits (全套服装) you could buy with $50. Then calculate and check your guess.

How might the cost of one week's clothes be different based on the weather?

Extension

Compare the cost of a trendy outfit you might buy from a department store to the cost of a school uniform. Based on what you find, do you think the argument about cost is a good one? Will the costs be different for boys and girls?

Standardized Test Preparation 2

标准考前测试 2

1 Which of these answers shows how to break down（分解） 5 x 14 into math facts you know?

Ⓐ (5 x 10) + (5 x 4)

Ⓑ (5 x 12) + (5 x 3)

Ⓒ (5 x 10) – (5 x 4)

Ⓓ (5 x 10) + (5 x 7)

2 What is the value of the 6 in 8,**6**72,913?

Ⓕ 6 hundreds

Ⓖ 6 ten thousands

Ⓗ 6 millions

Ⓙ 6 hundred thousands

3 Use the long division method to find the quotient.

479 ÷ 6 = ____

Ⓐ 89 R5

Ⓑ 77 R5

Ⓒ 79 R6

Ⓓ 79 R5

4 What is the product of 0.57 x 10?

Ⓕ 0.057

Ⓖ 57

Ⓗ 570

Ⓙ 5.7

5 Which of these numbers is a multiple（倍数） of 5 and a factor of 20?

Ⓐ 15

Ⓑ 10

Ⓒ 25

Ⓓ 12

6 Pick one question from this test. Explain how and why you chose your answer.

____

____

____

____

____

____

Methods of Dividing

除 法

Directions: Complete the problems below.

1. Complete the following problems using a method of division.

a. 123 ÷ 6 =	**b.** 502 ÷ 5 =	**c.** 321 ÷ 7 =
d. 827 ÷ 4 =	**e.** 285 ÷ 9 =	**f.** 417 ÷ 8 =

2. Use any method of dividing to complete the following problems.

a. 148 ÷ 8 =	**b.** 291 ÷ 9 =	**c.** 165 ÷ 3 =
d. 881 ÷ 7 =	**e.** 942 ÷ 6 =	**f.** 805 ÷ 3 =

Guessing and Checking Problems

估算与验算 练习

Directions: Solve each problem and show your work.

Problem A

Find three consecutive（连续的） 3-digit（3位） odd numbers（奇数） with a sum of 411.

Problem B

The school cafeteria（自助餐厅） served 78 lunches（午餐）. Pizza was the favorite choice selected by the greatest number of students. There was one-third the number of hamburgers（汉堡） as pizzas served. There were twice as many tacos（炸玉米饼） as hamburgers served. How many of each food item were selected by the students?

Standardized Test Preparation 3
标准考前测试 3

1 Which number is divisible by 2 and 5?

Ⓐ 31

Ⓑ 14

Ⓒ 40

Ⓓ 35

2 Which of the following answers is equal to $\frac{1}{10}$?

Ⓕ $\frac{7}{100}$

Ⓖ $\frac{1}{5}$

Ⓗ $\frac{11}{100}$

Ⓙ $\frac{10}{100}$

3 What is the next number in this sequence?

5, 16, 27, 38, ___

Ⓐ 48

Ⓑ 47

Ⓒ 49

Ⓓ 51

4 Use the long division method to find the quotient.

$7\overline{)367}$

Ⓕ 52 R4

Ⓖ 52 R3

Ⓗ 502 R3

Ⓙ 51 R3

5 Andy bought a green lizard（绿蜥蜴）for $7.95 and a small iguana（鬣蜥蜴）for $19.49. How much did he spend in all?

Ⓐ $28.54

Ⓑ $27.54

Ⓒ $26.44

Ⓓ $27.44

6 Pick one question from this test. Explain how and why you chose your answer.

____

____

____

____

____

____

Lesson 4

At the Fair
在游乐场

ENTRANCE（入场费）	
Adult	$3.75
Child (over 10)	$2.25
Child (under 10)	$1.25

SNACKS（快餐）	
Cold Drink（冷饮）	70¢
Hot Dog（热狗）	$1.10
Potato Chips（油炸薯片）	60¢
Ice Cream（冰激凌）	90¢
Chocolate Bar（巧克力棒）	45¢
Cookie（饼干）	65¢

RIDES（游乐设施）	
Roller Coaster（过山车）	$2.40
Waltzer（华尔兹）	$2.25
Big Wheel（摩天轮）	$2.05
Ghost Train（魔鬼火车）	$2.30
Carousel（旋转木马）	$2.10
Hook-a-Duck（钓鸭子）	$2.15

At the Fair *(cont.)*

Directions: Complete the following problems.

1. Calculate the total cost for each group to go to the fair.

Group A	Group B	Group C
4 adults	2 adults	12 adults
3 children (over 10)	6 children (over 10)	2 children (over 10)
2 children (under 10)	1 child (under 10)	
Total ______________	Total ______________	Total ______________

2. Janine, Malik, and Louise go to the fair. After paying their admission fee（门票费）($2.25 each), they each have $10 to spend. They go on two rides each. Janine has $5.85 left, Malik has $5.65, and Louise has $5.30. Which rides did each student go on?

3. How much of his $5 will Raul have after he has a ride on the Waltzer and buys a drink and ice cream?

4. Jane has $2.50. Can she afford to have a ride on the carousel and buy a hot dog?

5. How much would it cost Mariah to buy three drinks, three ice creams, four cookies, and two hot dogs? How much change would she have from $20?

6. Desmond has $10. He goes on three different rides and buys one bar of chocolate. He has $2.85 left. Which three rides did he go on?

7. How much would it cost for a family of four (two adults and two children over the age of 10) to go on each ride once in addition to（除……之外）the entrance fee（入场费）?

8. Plan a visit to the fair for two adults, a child over 10, and two children under 10. They each want to go on three rides and have something to eat and drink. How much would it cost? Show your plan for their visit.

Pony Express

快马邮递

In 1860, a letter mailed in Missouri（密苏里） would take 20 days to arrive in Sacramento（萨克拉门托）, California. Some people thought that was too long. William M. Gwin, Alexander Majors, and Daniel Phelps decided to do something about it. They started the Pony Express, a mail service using horses and riders（骑手）. They bought 400 very fast horses and hired 75 riders.

A rider could not weigh more than 125 pounds. The ages of riders ranged from 11 years old to 40 years old. The youngest rider may have been a boy named Bronco Charlie Miller. Each rider earned $100 a month, which was a lot of money at the time.

Six days a week, Monday through Saturday, a Pony Express rider left St. Joseph, Missouri, at noon. Another rider left Sacramento, California, at 8:00 in the morning. Each carried mail in a special saddlebag（鞍囊）. Each rode for 75 to 100 miles, changing horses along the way. Then, he handed the mail to another rider who rode for 75 to 100 miles. This relay continued until the end of the route. The Pony Express took only 10 days to get mail from Missouri to California, a distance of 1,966 miles.

Sending mail by Pony Express was costly（昂贵的）. A half-ounce letter cost $5.00. Each additional half-ounce cost $1.00.

The Pony Express went out of business（停业） in 1861, less than two years after it started. The telegraph helped bring an end to this clever solution for speeding up（加速） communication（通信）.

Pony Express *(cont.)*

What Is the Problem?

You are a Pony Express dispatcher（派遣员） in Missouri. You have to manage the mail, the horses, and the riders for the trip to California.

You will be asked to work on the problems below, using the guessing and checking strategy. Use the information below and on page 19 to answer the questions.

Before you begin solving each problem, be sure to locate the information you will need. Think about what information you can use to help guide your guesses. Then, you may want to organize your guesses into a table to keep track of them.

Remember: If you have trouble planning ways to solve each problem, reread it and rethink your strategy. When you have a solution, think about whether your answer makes sense.

Problem-Solving Strategy: Guessing and Checking

Problem A

The combined age of two Pony Express riders is 50. One rider is 20 years older than the other. How old is each rider?

Hint: Use the age range of riders to help guide your guesses.

Problem B

Two letters have a combined weight of six ounces. One letter weighs two ounces more than the other letter. How much does each letter weigh?

Using the information from the previous page, how much will each letter cost to send? How many ounces could you add to the smaller letter in order to spend less than $21.00 total for both letters?

Problem C

On a trip from Missouri to California, two riders ride a combined total of 170 miles. One rider rides 20 miles more than the other rider. How many miles does each rider ride?

Hint: Use the range of the number of miles a rider rode in a day to guide your guesses.

If these are the last two riders in the trip, how many miles were traveled before the mail got to them?

Extension

If riders rode an average of 90 miles each, how many riders would be needed to travel the distance from Missouri to California?

Standardized Test Preparation 4

标准考前测试 4

1 What is half of 0.8?

Ⓐ 1.6

Ⓑ 0.04

Ⓒ 0.4

Ⓓ 0.06

2 Which of these numbers is a multiple of 2 and a factor（因数） of 12?

Ⓕ 3

Ⓖ 4

Ⓗ 5

Ⓙ 10

3 Mount Everest is twenty-nine thousand, thirty-five feet tall. Which number shows this height?

Ⓐ 2,935

Ⓑ 29,350

Ⓒ 29,035

Ⓓ 290,035

4 Which is an example of perimeter?

Ⓕ the distance around a playground

Ⓖ the depth（深度） of a lake

Ⓗ the height of a building

Ⓙ the length of a straight line across a field

5 Which of the following numbers is a prime number（素数）?

Ⓐ 28

Ⓑ 215

Ⓒ 39

Ⓓ 71

6 Pick one question from this test. Explain how and why you chose your answer.

____

____

____

____

____

____

Lesson 5

In the Balance

找位置

Directions: Use the number line to determine the location of the point for each problem. Write your answer on the line below each star.

1.

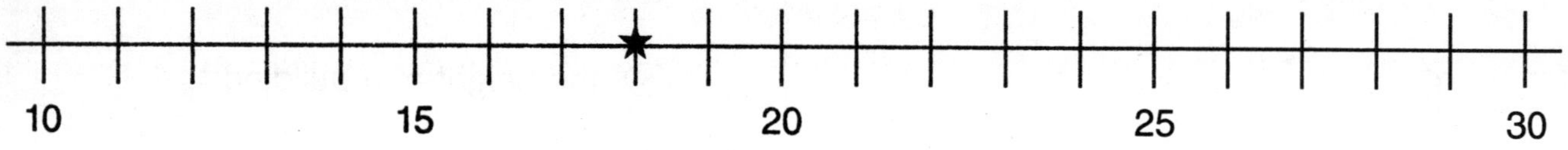

______

2.

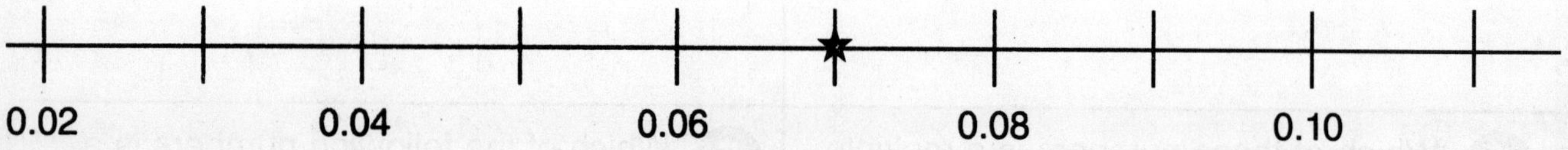

______

3.

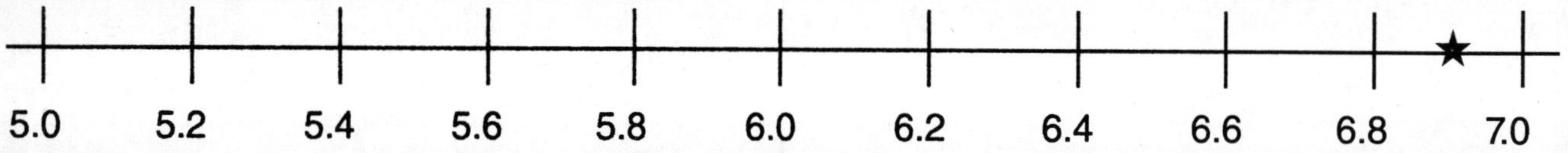

______

4.

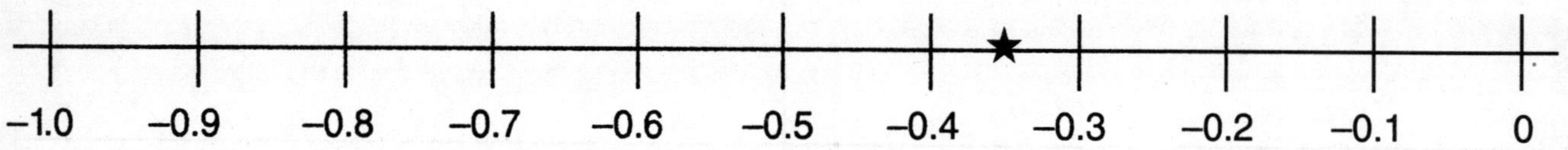

______

5.

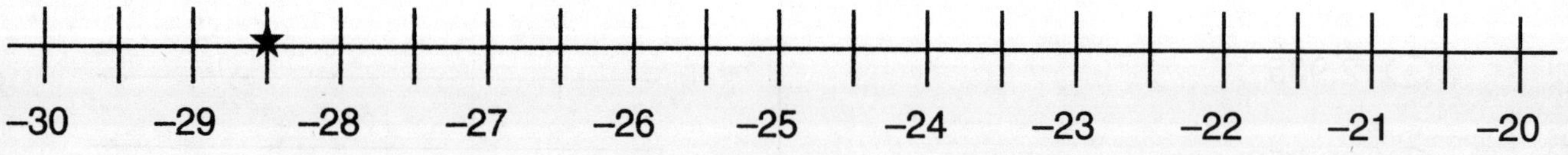

______

6.

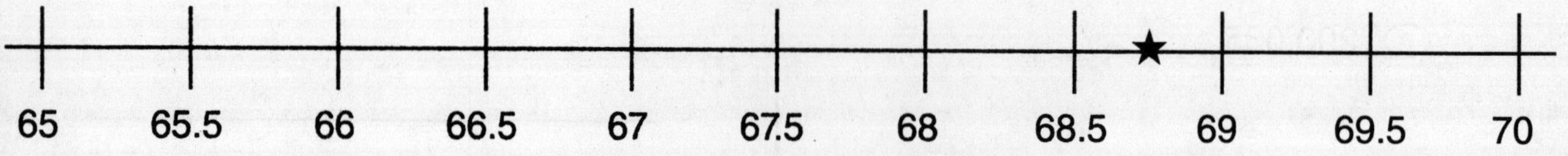

______

Decimal Cards

小数卡

Directions: Cut out these cards and place them in numerical order（数值次序）.

0.55	0.12
0.210	0.2
0.31	0.405
0.102	0.451
0.45	0.305

Decimal Cards *(cont.)*

This page is left blank so you can cut out the manipulatives（操作）*on the previous page*（前一页）.

Guessing and Checking Group Problems

估算与验算 练习

Directions: Solve Problem A and show your work. Using the strategy（方法） you've studied this week, write and solve your own problems for Problem B and Problem C.

Problem A

Jerome must make $0.60. He must use at least one of each coin（硬币）—penny（1美分）, nickel（5美分）, dime（10美分）, and quarter（25美分）—and no more than 10 coins in all. How many of each coin should he use?

Problem B: ____________________

____

____

____

____

____

Show work:

Answer: ____________________

Problem C: ____________________

____

____

____

____

____

Show work:

Answer: ____________________

Standardized Test Preparation 5

标准考前测试 5

1 At the County Fair, Cheryl spent \$2.35 on popcorn（爆米花） and \$1.89 for a lemonade（柠檬水）. What was her total cost?

Ⓐ \$4.24

Ⓑ \$4.34

Ⓒ \$3.94

Ⓓ \$4.26

2 Solve（求解）.

7 x 17 = ____

Ⓕ 149

Ⓖ 111

Ⓗ 119

Ⓙ 121

3 Determine the next 3 numbers in this sequence.

1, 6, 11, 16, ___, ___, ___

Ⓐ 21, 26, 31,

Ⓑ 21, 25, 30

Ⓒ 21, 23, 27

Ⓓ 26, 36, 46

4 The art teacher used a blank canvas（空白画布） for her painting. What is the area of the canvas if its length is 6 inches and its width is 4 inches?

Ⓕ 24 $in.^2$

Ⓖ 12 $in.^2$

Ⓗ 48 $in.^2$

Ⓙ 16 $in.^2$

5 What is the approximate length of a new pencil?

Ⓐ 7 feet

Ⓑ 7 inches

Ⓒ 7 centimeters

Ⓓ 7 yards

6 Pick one question from this test. Explain how and why you chose your answer.

____

____

____

____

____

____

Lesson 6

Code Wheels
代码转盘

Directions: Solve the problems below. Convert（转换） each answer into a letter from the code wheels.

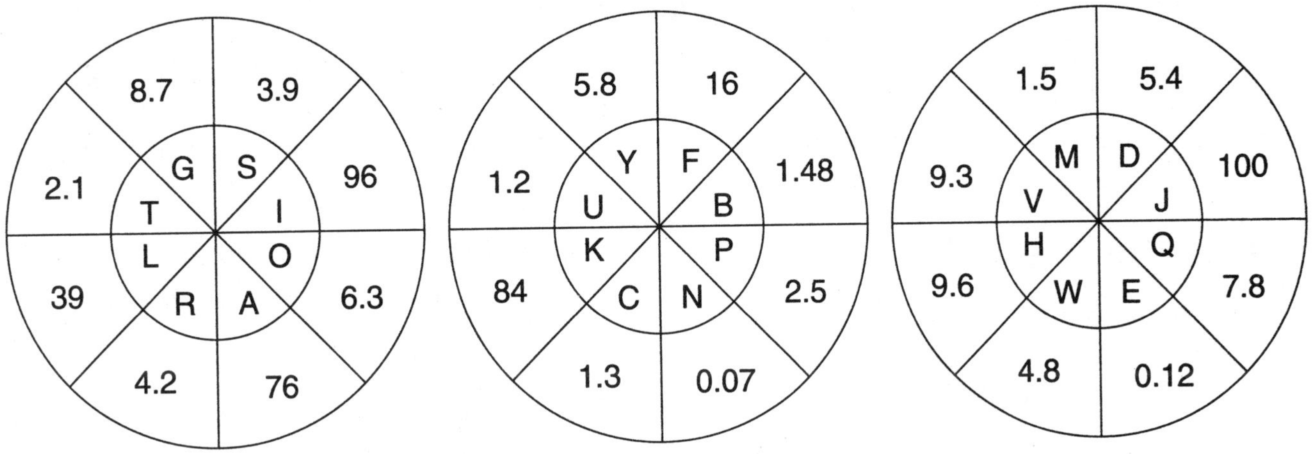

1. 9.6 x 10 ________
2. 7 ÷ 100 ________
3. 0.8 x 6 ________
4. 8 x 1.2 ________
5. 960 ÷ 10 ________
6. 0.65 x 2 ________
7. 0.96 x 10 ________
8. 0.75 x 2 ________
9. 7 x 0.9 ________
10. 0.14 ÷ 2 ________
11. 7 x 0.3 ________
12. Double 4.8 ________
13. 6 x 16 ________
14. 39 ÷ 10 ________
15. Double 2.9 ________
16. 0.63 x 10 ________
17. 0.3 x 4 ________
18. 12.6 ÷ 3 ________
19. 0.74 x 2 ________
20. 10 x (Double 4.8) ________
21. Halve 8.4 ________
22. 10 x 0.21 ________
23. 19.2 ÷ 2 ________
24. 54 ÷ 10 ________
25. 0.76 x 100 ________
26. 10 x 0.58 ________

What is the coded message?

___ ___   ___ ___ ___ ___ ___   ___ ___ ___ ___ ___   ___ ___

___ ___ ___ ___   ___ ___ ___ ___ ___ ___ ___ ___?

Creating a Table

制图表

When you have to solve a problem, it is very helpful to place the information in a table. A table organizes the given information so that you understand what the problem is telling you and what you still have to find out in order to solve the problem.

Using a table can help you keep track of the information and reduce the chance of making mistakes. A table makes it easy to see what information is given and what information is missing. When a table is created, the information often shows a pattern, which can help you solve the problem. With practice, you will learn how to use a table to solve problems.

Planning Your Table

- First, read the problem and decide what information is given and what information you need to find out.
- Next, be sure you understand what the table is going to tell you. Include headings to label the information that you are organizing.

Problem: Poetry Reading (诗词朗诵)

The Problem

A group of students is learning a long poem for a school assembly (集会). Each week, they are taught an increasing number of verses (诗篇). The first week they learn 1 verse, and at the end of the second week, they can say 3 verses. At the end of the third week, the students can recite (背诵) 6 verses. At the end of the fourth week, they know 10. Continuing with this pattern, how many verses would they be able to recite after 8 weeks?

Understanding the Problem

- *What do we know?*

  In the first week, students learn 1 verse. At the end of the second week, they know 3 verses. At the end of the third week, they know 6 verses. By the end of the fourth week, they know 10 verses.

- *What do we need to find out?*

  How many verses they know after 8 weeks.

Planning and Communicating a Solution

To solve this problem, create a table with 2 rows and 9 columns (or 2 columns and 9 rows).

The first row should list the week numbers (1–8). The second row should list the increasing number of verses.

Place the known numbers in the table, and identify the pattern. Complete the table, using the pattern.

Week	1	2	3	4	5	6	7	8
Number of Verses	1	3	6	10	15	21	28	36

- *Do you see the answer?*

  The pattern for the number of verses is +2, +3, +4, +5, ... The students would be able to recite 36 verses by the end of week 8.

Reflecting and Generalizing

The table makes it easy to see a pattern. If you discover the pattern right away, you can simply complete the pattern mentally to solve the problem.

Extension

How would it change the results if the students did not practice for one week?

Standardized Test Preparation 6

标准考前测试 6

1 An apple tree is 31 feet tall. Which of the following is the closest measurement?

Ⓐ 12 yd.

Ⓑ 10 yd. 12 in.

Ⓒ 28 ft. 11 in.

Ⓓ 32 ft. 1 in.

2 Alex bought a goldfish（金鱼） on sale for $2.99. The original price was $4.25. How much did he save from the original price?

Ⓕ $1.26

Ⓖ $2.26

Ⓗ $1.24

Ⓙ $2.06

3 What is the value of 6 in 90**6**,415?

Ⓐ 6 ten thousands

Ⓑ 6 millions

Ⓒ 6 thousands

Ⓓ 6 hundreds

4 What is the product of 0.97 x 100?

Ⓕ 9.7

Ⓖ 970

Ⓗ 97

Ⓙ 0.097

5 Simplify $\frac{81}{90}$ by finding the greatest common factor.

Ⓐ $\frac{3}{4}$

Ⓑ $\frac{7}{10}$

Ⓒ $\frac{9}{10}$

Ⓓ $\frac{5}{10}$

6 Pick one question from this test. Explain how and why you chose your answer.

____

____

____

____

____

____

Lesson 7

It's a Fact!
这才是真相!

Directions: Complete the following problems. Show your work on a separate sheet of paper.

1. Find all the factors (因数) of these numbers.

   a. 16 ______ c. 55 ______

   b. 30 ______ d. 100 ______

2. Find the factors that these two numbers have in common (共同的).

   a. 20 and 30 ______ c. 15 and 45 ______

   b. 49 and 28 ______ d. 10 and 36 ______

3. Find the largest factor that these two numbers have in common.

   a. 10 and 15 ______ c. 6 and 21 ______

   b. 12 and 48 ______ d. 35 and 14 ______

4. Mark a dot (点) on the graph below where each of the numbers from 1–24 has a factor. The number 5 has been done for you. It is a prime number, so the only factors are 1 and 5.

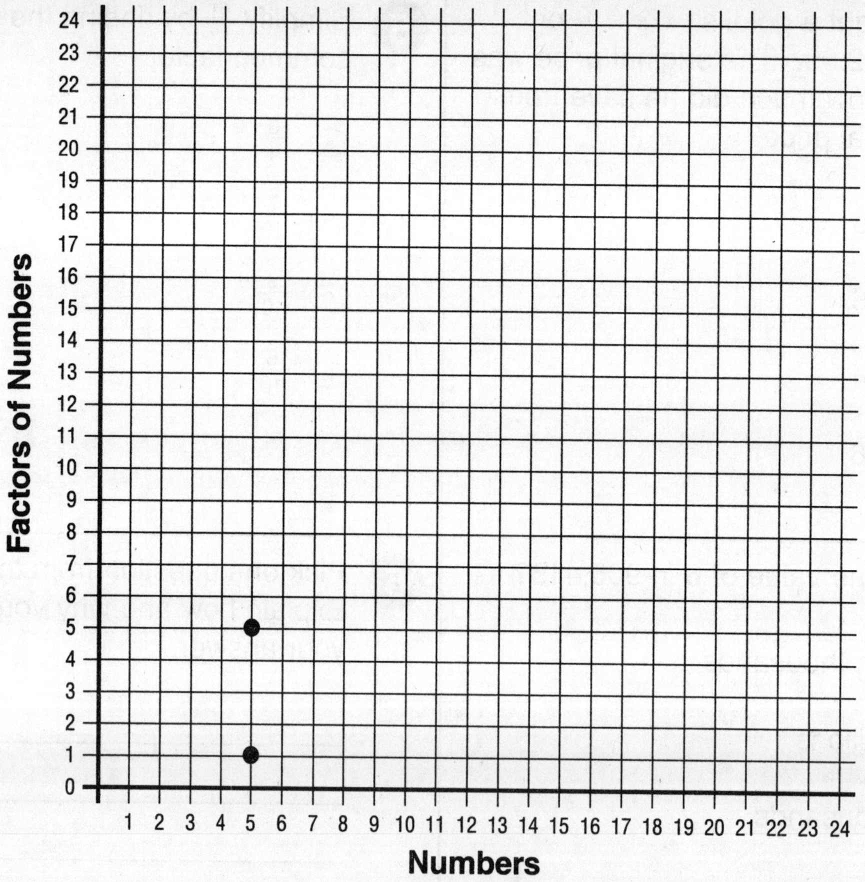

   a. Describe the completed graph. Can you see any patterns? ______

   ______

   b. Which number has the most factors? ______

   c. Use the graph to find the common factors for 8 and 12. ______

   d. Find the greatest common factor of 12 and 24. ______

Chad's Movie Rental Store
查德的影碟出租店

Do you want to see a movie? Maybe you can rent（租借）a DVD. There are places called movie rental stores that allow you pay money to borrow movies. You have to bring the movies back when you're done with them.

Movie rental stores have every kind of movie, from comedies（喜剧片）to action films（动作片）. Sometimes there are video games（电子游戏）to rent, as well.

In the United States, movie rentals make over 10 billion dollars（100亿美元）a year! That is a lot of money! Think of how many DVDs must be rented in a year.

Every movie rental store has its own（自己的）rules. Sometimes new movies can only be rented for a day or two. Sometimes old movies can be kept for a week or longer. The cost to rent a new movie might be $3.00 or more. Most stores charge a fine for the late return of a DVD.

Chad's Movie Rental Store has these rules for renting movies:

1. **New Movie Rentals $3.00**
2. **Older Movie Rentals $2.00**
3. **A $1.00 fine is charged for each day a movie is late.**
4. **A new movie must be returned by ten o'clock in the morning on the third day of the rental.**
5. **An older movie is due back by ten o'clock in the morning on the seventh day of the rental.**

**Keep in mind that the day you rented the movie counts as the first day of rental. Remembering this will help you figure out when a movie must be returned.*

Chad's Movie Rental Store *(cont.)*

What Is the Problem?

Imagine you are renting movies form Chad's Movie Rental Store. How much will it cost you to rent movies?

Use the information on page 31 and in the problems below to create a table and answer the questions.

Before you begin solving each problem, think about what information is given and what information you will need to figure out in order to complete the problem. Think about how you will create your table and what you will include in it.

Remember: If you have trouble planning ways to solve each problem, reread it and rethink your strategy. When you have a solution, think about whether your answer makes sense.

Problem-Solving Strategy: Creating a Table

Problem A

Create a table to show when new movies must be returned, depending on what day they were rented. Make a table with two columns. Label one column *Day Movie Is Rented* and the other column *Day New Movie Is Due*. Write the days of the week in the first column. Complete the second column to show the days the new movies are due.

Problem B

You rent one old movie and one new movie on Friday. You return both movies on Saturday. How much is the rental fee? Do you pay a fine?

Add a third column to the table you created in Problem A. In this column, figure out what day old movies have to be returned.

Problem C

You rent two new movies and one old movie on Friday. How much is the rental fee? When are the movies due back to Chad's Movie Rental Store? If you return the new movies on Tuesday, will you have to pay a late fee（滞纳金）? If so, how much do you pay? How much did you pay in rental fees and fines for the three movies?

Extension

Imagine how you would manage your own movie rental store. Create a table that lists rental due dates and prices for renting movies. Figure out cost and due dates（付款日）. Will you have different due dates for new movies and old movies? If so, show them in your table. Determine the cost of renting a movie and the fine that will be charged for overdue（过期的）movies. If old movies cost less than new ones, include that in the table.

Standardized Test Preparation 7

标准考前测试 7

1 What is the volume of the shape?

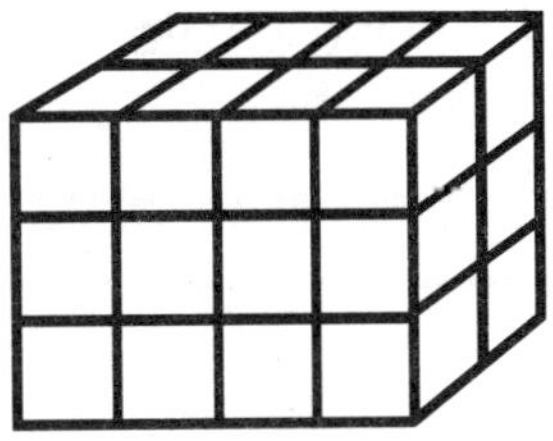

Ⓐ 24 units squared

Ⓑ 12 units cubed

Ⓒ 24 units cubed

Ⓓ 12 units squared

2 Which number can be divided evenly（整除） into 51?

Ⓕ 5

Ⓖ 14

Ⓗ 3

Ⓙ 7

3 Solve.

25 x 12 = ___

Ⓐ 350

Ⓑ 250

Ⓒ 275

Ⓓ 300

4 A dinner casserole（盘） of macaroni（通心粉） and cheese（奶酪） weighed 62 oz. Which of the following is the closest measurement?

Ⓕ 4 lb.

Ⓖ 3 lb. 14 oz.

Ⓗ 70 oz.

Ⓙ 4 lb. 4 oz.

5 Which shape fits in area C of this Venn diagram?

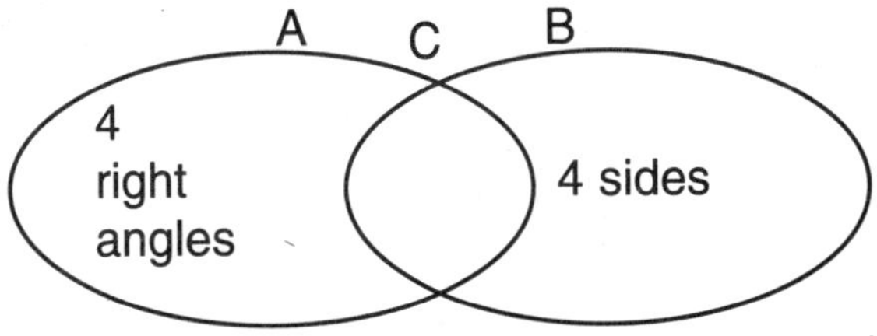

Ⓐ rectangle（矩形）

Ⓑ rhombus（菱形）

Ⓒ triangle（三角形）

Ⓓ trapezoid（梯形）

6 Pick one question from this test. Explain how and why you chose your answer.

____

____

____

____

____

____

Hundred Chart

百位数表

Directions: Shade the numbers that are divisible by 2, 3, 4, and 5. Use a different color for each set of numbers.

1	2	3	4	5	6	7	8	9	10
11	12	13	14	15	16	17	18	19	20
21	22	23	24	25	26	27	28	29	30
31	32	33	34	35	36	37	38	39	40
41	42	43	44	45	46	47	48	49	50
51	52	53	54	55	56	57	58	59	60
61	62	63	64	65	66	67	68	69	70
71	72	73	74	75	76	77	78	79	80
81	82	83	84	85	86	87	88	89	90
91	92	93	94	95	96	97	98	99	100

The Sieve of Eratosthenes
爱拉托逊斯筛法

Eratosthenes of Cyrene（昔兰尼）(276–194 b.c.) devised（设计） a "sieve（筛法）" to identify prime numbers. A sieve is like a strainer（过滤网） that you can use to drain spaghetti（沥干意大利面） when it is finished cooking. The water drains out, leaving spaghetti behind. The Sieve of Eratosthenes drains out composite numbers（复合数） and leaves prime numbers behind.

Directions: You can find all of the prime numbers from 1 to 100 using the Sieve of Eratosthenes. Use a marker（记号笔）, crayon（蜡笔）, or pencil（铅笔） to do the following:

- Color the number 1. It has only one factor (1 x 1) so it is not prime.
- Color all of the multiples of 2, except the number 2.
- Color all of the multiples of 3, except the number 3.
- Color all of the multiples of 5, except the number 5.
- Color all of the multiples of 7, except the number 7.

When finished coloring, make a list of all the numbers that have not been colored. These are called prime numbers because they have no factors other than 1 and themselves.

1	2	3	4	5	6	7	8	9	10
11	12	13	14	15	16	17	18	19	20
21	22	23	24	25	26	27	28	29	30
31	32	33	34	35	36	37	38	39	40
41	42	43	44	45	46	47	48	49	50
51	52	53	54	55	56	57	58	59	60
61	62	63	64	65	66	67	68	69	70
71	72	73	74	75	76	77	78	79	80
81	82	83	84	85	86	87	88	89	90
91	92	93	94	95	96	97	98	99	100

Creating a Table Problems

制图表 练习

Directions: Solve each problem and show your work.

Problem A

In a bag, there are 10 triangles and squares altogether. There are more triangles than squares. There are 31 sides in all. How many of each shape are in the bag?

Problem B

Look at each table. Write the rule.

A.

In	Out
1,059	1,064
2,541	2,546
2,910	2,915
6,701	6,706
1,037	1,042

B.

In	Out
2,356	2,355
4,762	4,761
8,254	8,253
3,889	3,888
3,268	3,267

A. ____________

B. ____________

C.

In	Out
1,026	1,070
1,217	1,261
3,953	3,997
8,853	8,897
1,644	1,688

D.

In	Out
3,611	3,600
9,791	9,780
7,897	7,886
9,657	9,646
4,180	4,169

C. ____________

D. ____________

Standardized Test Preparation 8

标准考前测试 8

1 Determine the next two numbers in this sequence.

9, 14, 20, 27, 35, ___, ___

Ⓐ 40, 47

Ⓑ 43, 56

Ⓒ 42, 50

Ⓓ 44, 54

4 What is the total area of the front and back of the card?

Ⓕ 14 in.^{2}

Ⓖ 40 in.^{2}

Ⓗ 48 in.^{2}

Ⓙ 24 in.^{2}

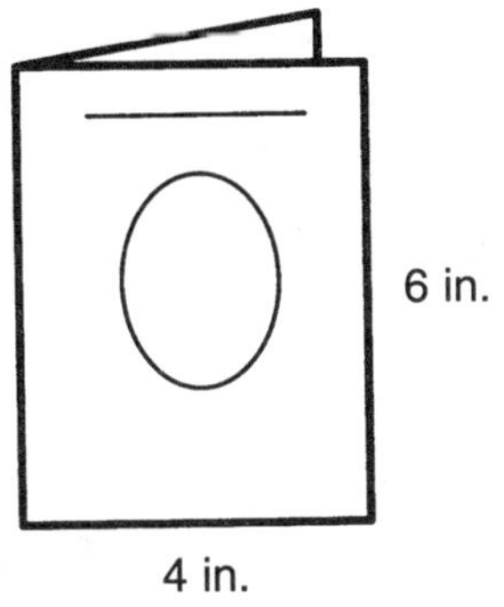

2 Which of these numbers is a factor of both 12 and 28?

Ⓕ 4

Ⓖ 6

Ⓗ 8

Ⓙ 7

5 What number means seven millions, no hundred thousands, four ten thousands, three thousands, no hundreds, no tens, and five ones?

Ⓐ 704,315

Ⓑ 7,043,050

Ⓒ 7,043,005

Ⓓ 7,430,005

3 At the County Fair, C.J. spent \$3.89 for a hamburger and \$2.29 for a soft drink. What was the total cost?

Ⓐ \$6.17

Ⓑ \$6.18

Ⓒ \$6.28

Ⓓ \$5.18

6 Pick one question from this test. Explain how and why you chose your answer.

____

____

____

____

____

____

Fraction Wall
分数墙

Directions: Use the fraction wall to answer the follow questions.

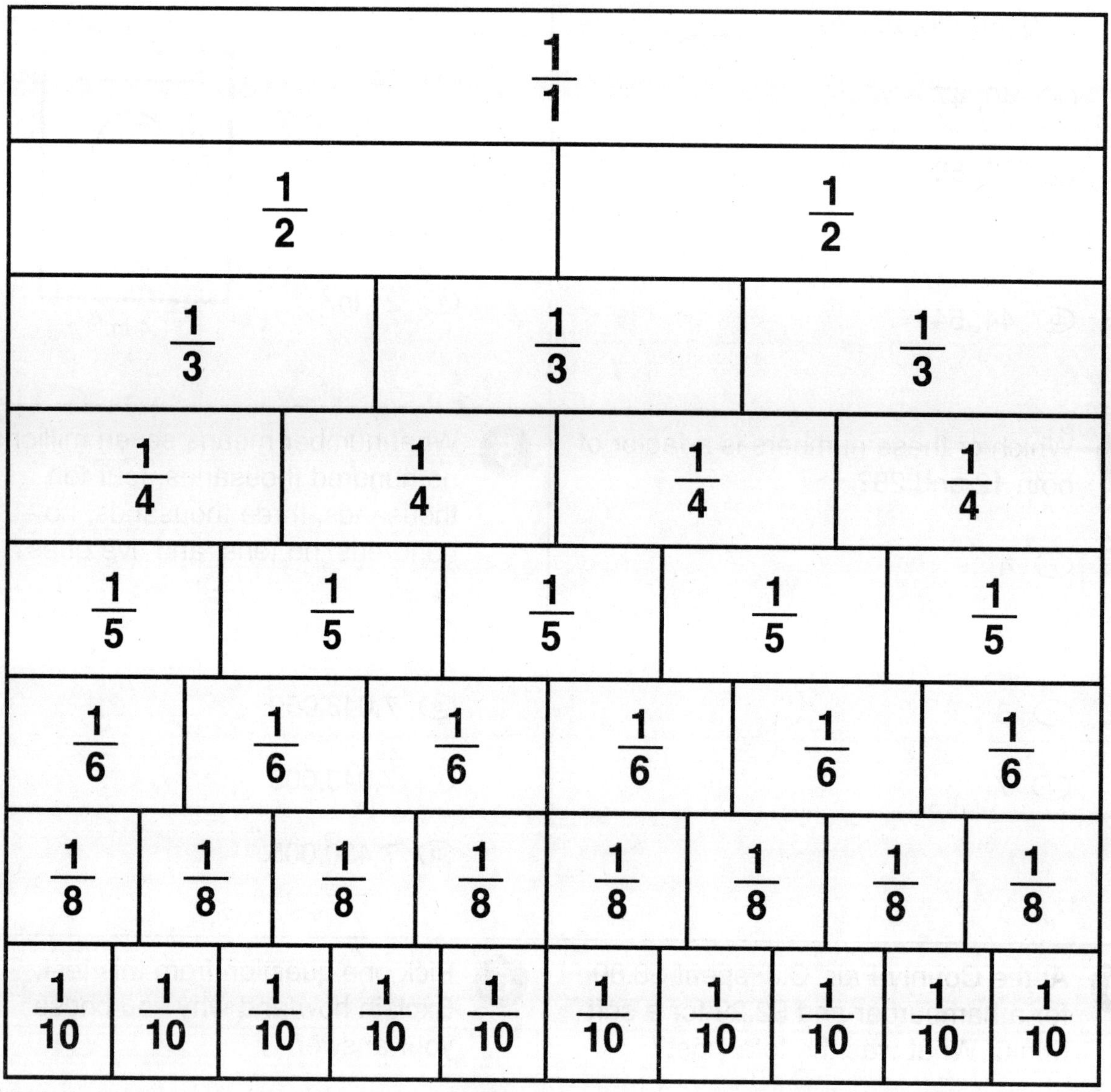

How many one-fourth pieces do you see under the one whole?

What fraction respresents the 3 one-eighth pieces?

How many one-tenth pieces equal two-fifths?

Adding and Subtracting Fractions
分数加减法

Directions: Complete the addition and subtraction problems below. Simplify if possible.

1. $\frac{2}{10} + \frac{7}{10} =$ ______________

2. $\frac{5}{12} - \frac{2}{12} =$ ______________

3. $\frac{9}{10} - \frac{7}{10} =$ ______________

4. $\frac{6}{8} + \frac{1}{8} =$ ______________

5. $\frac{10}{13} + \frac{2}{13} =$ ______________

6. $\frac{5}{7} - \frac{4}{7} =$ ______________

7. $\frac{2}{6} + \frac{3}{6} =$ ______________

8. $\frac{8}{11} - \frac{3}{11} =$ ______________

9. $\frac{4}{3} - \frac{1}{3} =$ ______________

10. $\frac{9}{13} + \frac{1}{13} =$ ______________

Directions: Answer the questions below using the charts.

For a science project（科学项目）, Mrs. Gaston's science class is recording the growth of their blue iguana（蓝岩鬣蜥）at the end of every month. Her class uses a table like this:

Month（月份）	Growth in Inches
September（9月）	$\frac{1}{8}$ inch
October（10月）	$\frac{3}{8}$ inch
November（11月）	$\frac{5}{8}$ inch
December（12月）	$\frac{1}{8}$ inch

11. How many inches had the blue iguana grown by the end of October?

____

12. How many inches had the blue iguana grown by the end of December?

____

13. If the blue iguana was $3\frac{3}{8}$ inches long by the end of September, how long was the iguana at the end of August?

____

For a science fair project, Javier is recording the amount of water that evaporates（蒸发）from a bucket（桶）in a month. He creates a table like this:

Week	Amount of Water that Evaporates
1	$\frac{2}{16}$ inch
2	$\frac{1}{16}$ inch
3	$\frac{3}{16}$ inch
4	$\frac{2}{16}$ inch

14. How many inches of water had evaporated from the bucket at the end of Week 2?

____

15. How many total inches of water evaporated?

____

16. If Javier originally put 4 inches of water in the bucket, how many inches of water were left after the experiment was completed? ______________

The Popularity of Pizza

比萨的普及

Mario's Pizza Palace

Toppings (馅料)

Pepperoni (意大利辣香肠)	$1.50
Mushroom (蘑菇)	$1.25
Sausage (香肠)	$1.50
Extra Cheese (加量乳酪)	$0.75
Onions (洋葱)	$1.00
Anchovies (凤尾鱼)	$1.25
The Works (制作费)	$3.00

Prices for Cheese Pizza

Small (小号)	$8.00
Medium (中号)	$9.50
Large (大号)	$11.00
Extra Large (特大号)	$13.00

Pizza has been a favorite food for many years. Pizza got its start in ancient times. People in the Middle East (中东) cooked flat (扁平), round bread. Later, the Greeks (希腊人) and Romans (罗马人) added oil (油) and spices (调料) to the bread.

Over 200 years ago, an Italian baker (意大利面包师) was asked to make a new food. The food would be served to a visiting monarchs (君主). The baker wanted his food to have the colors of the Italian flag. He needed something that had red, white, and green in it. So the baker put red sauce (红调味汁), white cheese, and green spices on flat bread. This was the first official (正式的) pizza. The king and queen (王后) liked it a lot. Word spread, and soon everyone wanted to try this new food.

Today, people around the world enjoy pizza. Many companies make hundreds of pizzas each day. Many places serve pizza, and people can order it and take it home. Pizza is a big business.

The Popularity of Pizza *(cont.)*

What Is the Problem?

You like to eat the pizza from Mario's Pizza Palace. When you go there, there are many choices to make. You need to choose the size of the pizza and the toppings. Take a trip to Mario's and place your order.

Use the information on page 40 and below to create a table and answer the questions.

Before you begin solving each problem, think about what information is given and what information you will need to figure out in order to complete the problem. Think about how you will create your table and what you will include in it.

Remember: If you have trouble planning ways to solve each problem, reread it and rethink your strategy. When you have a solution, think about whether your answer makes sense.

Problem-Solving Strategy: Creating a Table

Problem A

What are your two top choices for sizes of pizza? What are your two favorite toppings? Your mom says you can order a pizza with one topping. How many different pizzas can you make with your choices for toppings and size? How much would each pizza cost? Which pizza would you order?

Problem B

If you are told that you can have two toppings on your pizza, create a table that shows your two top choices for two toppings and sizes.

How much would each pizza cost? Which pizza would you order?

Problem C

You want a pizza with two toppings. If you only have $15.00, what size pizzas with two toppings can you buy? Are there any size pizzas with two toppings that you cannot buy? If so, how much more money would you need?

Extension

As a class, make a table of everyone's top choices of toppings. A large pizza has eight slices. If you decide to order large pizzas with the most popular topping, how much would each pizza cost? If each student has two slices, how many pizzas would you need? How much would your class spend on pizza?

Standardized Test Preparation 9

标准考前测试 9

1 You would need to know the perimeter to tell _____.

Ⓐ the height of a picture frame（画框）

Ⓑ the length of a train track（火车轨道） through a station

Ⓒ the distance around a classroom

Ⓓ the weight of a notebook

2 Marissa has 396 stamps（邮票） in her stamp collection. How many stamps can she place on each page if she has 9 pages?

Ⓕ 404 stamps

Ⓖ 44 stamps

Ⓗ 46 stamps

Ⓙ 49 stamps

3 What are coordinates（坐标） for point C?

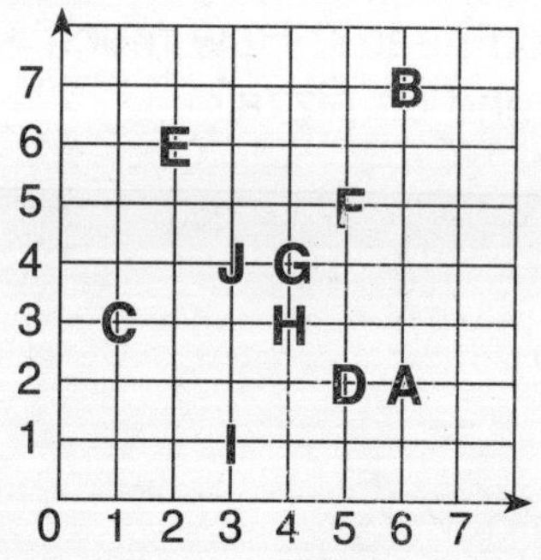

Ⓐ (3, 1)　　Ⓒ (3, 3)

Ⓑ (1, 3)　　Ⓓ (0, 3)

4 What is the approximate width of an individual student desk?

Ⓕ 10 inches

Ⓖ 2 feet

Ⓗ 2 yards

Ⓙ 8 feet

5 Which two lines are parallel?

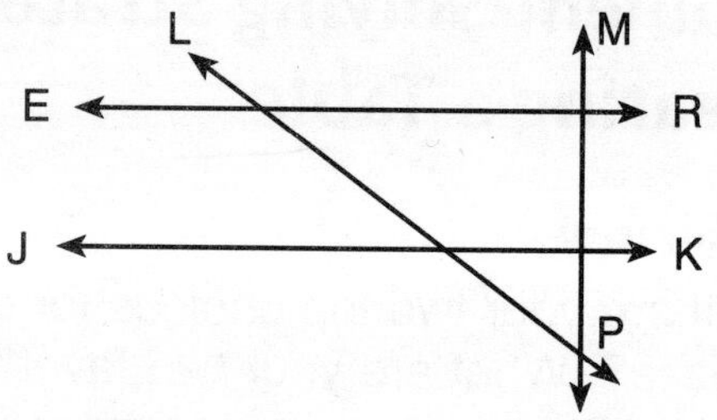

Ⓐ $\overleftrightarrow{PM}$ and $\overleftrightarrow{JK}$　　Ⓒ $\overleftrightarrow{JK}$ and $\overleftrightarrow{PM}$

Ⓑ $\overleftrightarrow{ER}$ and $\overleftrightarrow{JK}$　　Ⓓ $\overleftrightarrow{MP}$ and $\overleftrightarrow{PL}$

6 Pick one question from this test. Explain how and why you chose your answer.

____

____

____

____

____

____

Simply Fractions

约 分

Directions: Simplify these fractions by finding the greatest common factors.

1. $\frac{14}{21} =$

2. $\frac{8}{64} =$

3. $\frac{3}{9} =$

4. $\frac{16}{20} =$

5. $\frac{14}{84} =$

6. $\frac{10}{120} =$

7. $\frac{37}{148} =$

8. $\frac{24}{108} =$

9. $\frac{75}{100} =$

10. $\frac{15}{75} =$

11. $\frac{19}{133} =$

12. $\frac{275}{1025} =$

Challenge

Make up 10 of your own fractions for a classmate to simplify. Make sure that the numerator（分子）and denominator（分母）have a common factor.

Creating a Table Group Problems

制图表 分组练习

Directions: Solve Problem A and show your work. Using the strategy you've studied this week, write and solve your own problems for Problem B and Problem C.

Problem A

The Jelly Bean Machine produces jelly beans（软心豆糖）in the following sequence—5 blue, 6 red, and 9 yellow. After the Jelly Bean Machine has made 100 jelly beans, how many jelly beans of each color are there?

Problem B: ______________________

____

____

____

____

____

Show work:

Answer: ______________________

Problem C: ______________________

____

____

____

____

____

Show work:

Answer: ______________________

Standardized Test Preparation 10

标准考前测试 10

1 What is the name of a triangle with three equal sides?

Ⓐ equilateral（等边）

Ⓑ isosceles（等腰）

Ⓒ right（直角）

Ⓓ scalene（不等边）

2 Gerard received these grades on his spelling tests（拼写测试）: 90, 60, 55, 80, 70, 60, 75, 75, 60. Determine the median（中间值）.

Ⓕ 60

Ⓖ 80

Ⓗ 90

Ⓙ 70

3 Choose the best category for this statement: *You will travel to Venus*（金星）*tomorrow.*

Ⓐ Certain（一定）

Ⓑ Likely（很可能）

Ⓒ Unlikely（不太可能）

Ⓓ Impossible（不可能）

4 If a square has sides that are 7 cm long, what is its perimeter?

Ⓕ 28 cm

Ⓖ 21 cm

Ⓗ 14 cm

Ⓙ 7 cm

5 On which day did Jack sell the fewest pizzas?

Pizzas Sold at Jack's

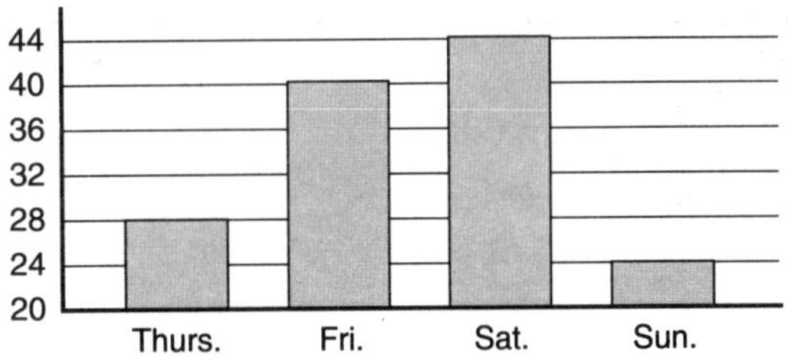

Ⓐ Thursday（周四）

Ⓑ Friday（周五）

Ⓒ Saturday（周六）

Ⓓ Sunday（周日）

6 Pick one question from this test. Explain how and why you chose your answer.

____

____

____

____

____

____

Improper Fractions

假分数

Directions: Simplify the improper fractions（假分数）below. Then write them as mixed numbers（带分数）.

1. $\frac{3}{2}$ = ____________ 6. $\frac{7}{5}$ = ____________

2. $\frac{8}{4}$ = ____________ 7. $\frac{19}{6}$ = ____________

3. $\frac{12}{10}$ = ____________ 8. $\frac{17}{10}$ = ____________

4. $\frac{100}{4}$ = ____________ 9. $\frac{10}{3}$ = ____________

5. $\frac{17}{14}$ = ____________ 10. $\frac{15}{2}$ = ____________

Directions: Write these mixed numbers as improper fractions.

11. $2\frac{1}{2}$ = ____________ 16. $11\frac{1}{8}$ = ____________

12. $4\frac{1}{8}$ = ____________ 17. $9\frac{4}{7}$ = ____________

13. $10\frac{1}{3}$ = ____________ 18. $3\frac{3}{10}$ = ____________

14. $7\frac{1}{6}$ = ____________ 19. $10\frac{3}{4}$ = ____________

15. $1\frac{2}{3}$ = ____________ 20. $5\frac{2}{5}$ = ____________

Challenge

Write five improper fractions. Exchange papers with a classmate and simplify the fractions.

1. ____________ = ____________
2. ____________ = ____________
3. ____________ = ____________
4. ____________ = ____________
5. ____________ = ____________

Looking for a Pattern
找规律

Mathematical patterns can be found everywhere—in numbers, shapes, and colorful combinations, for example. Therefore (因此), it is not surprising that looking for a pattern is a helpful strategy when you are trying to solve math problems. Once you find the pattern, it becomes easy to predict (预测) what comes next.

You will find that the problem-solving strategies you learn are sometimes connected to one another. You may use more than one strategy to solve a problem. As you learn more about "Looking for a Pattern," you will find that it is often an extension of "Creating a Table" or "Creating an Organized List."

Finding a Pattern in a Table

With some problems, you may need to create tables to easily look for patterns. You need to decide if there are 1, 2, 3, or more things that change (variables) in the pattern. Is a "total" column necessary (必要的)?

Problem: Apple Picking (摘苹果)

The Problem

When Janie went apple picking, 3 out of every 7 apples had wormholes (虫洞). How many good apples were there out of 49 total apples picked?

Understand the Problem

- *What do we know?*

  For every 7 apples picked, 4 are good and 3 have wormholes.

- *What do we need to know?*

  How many of the 49 apples that Janie picked are good.

Planning and Communicating a Solution

Solve the problem by creating a table with three columns labeled *Good Apples*, *Bad Apples*, and *Total Apples*.

You can see a pattern start to emerge (出现) in the table.

The *Good* column is increasing by 4 and the *Bad* column is increasing by 3. The *Total* column increases by 7.

Continue working with the pattern until you have 49 total apples.

Good Apples	Bad Apples	Total Apples
4	3	7
8	6	14
12	9	21
16	12	28
20	15	35
24	18	42
28	21	49

- *Do you see the answer?*

  Out of the 49 total picked apples, 28 apples will be good.

Reflecting and Generalizing

Once the table has been drawn, it is easy to see a pattern. The table helped to organize the information so that it could be seen clearly.

Extension

Jan needs 36 apples to share with her class. How many total apples will she have to collect to have 36 good apples?

Standardized Test Preparation 11
标准考前测试 11

1 Which of these units of measurement represents the best unit for measuring an earthworm（蚯蚓）?

Ⓐ meter（米）

Ⓑ foot（英尺）

Ⓒ yard（码）

Ⓓ inch（英寸）

2 Use mental math（心算） to solve this equation.

7 x 28 = ___

Ⓕ 149

Ⓖ 175

Ⓗ 196

Ⓙ 210

3 What words best describe these two figures（图形）?

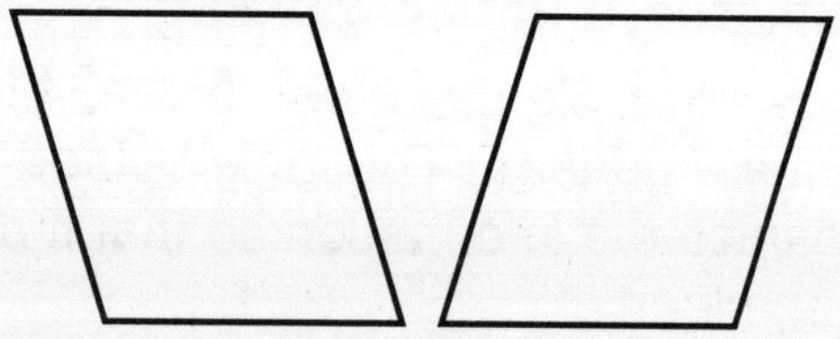

Ⓐ similar（相似）

Ⓑ congruent（全等）

Ⓒ quadrilateral（四边形）

Ⓓ scalene（不等边三角形）

4 What is the area of the rectangle?

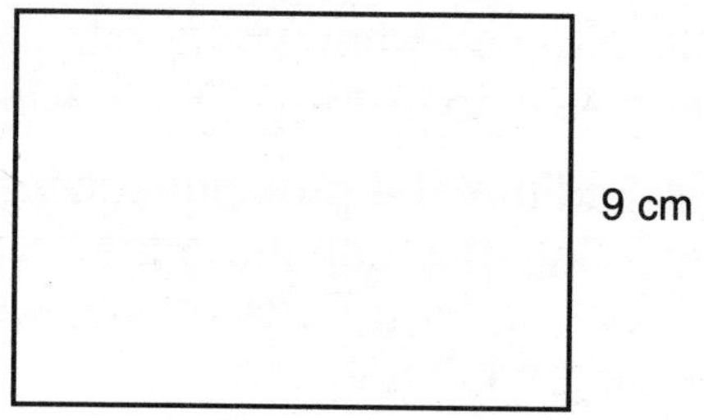

Ⓕ 121 cm^2　　Ⓗ 117 cm^2

Ⓖ 122 cm^2　　Ⓙ 147 cm^2

5 What are the coordinates for point A?

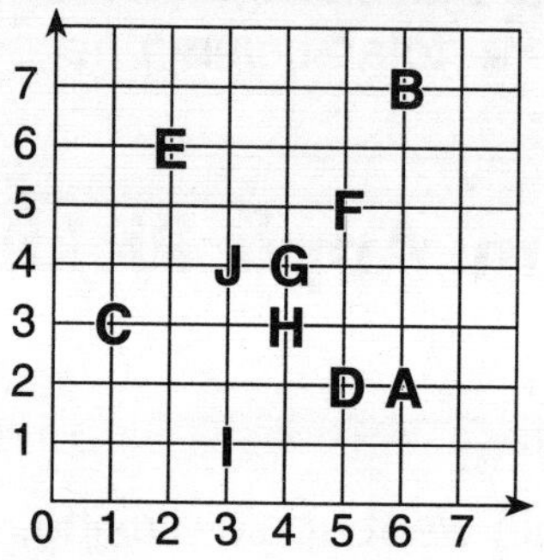

Ⓐ (6, 3)　　Ⓒ (6, 2)

Ⓑ (2, 6)　　Ⓓ (1, 7)

6 Pick one question from this test. Explain how and why you chose your answer.

____

____

____

____

____

____

On and On . . .

接下去！

Directions: Complete each of the following number sequences（数序）. Write the rule for each number sequence.

1. ______, ______, ______, 11, 14, 17, ______, ______, ______

   Rule: ______________________________

2. ______, ______, 9, 27, 81, ______, ______, ______

   Rule: ______________________________

3. 2, 10, 50, ______, ______, ______, ______

   Rule: ______________________________

4. 27, 23, 19, 15, ______, ______, ______, ______

   Rule: ______________________________

5. 1, 2, 4, 8, 16, 32, ______, ______, ______

   Rule: ______________________________

6. 1,024, 512, 256, 128, ______, ______, ______, ______

   Rule: ______________________________

7. 3, 5, 8, 12, ______, ______, ______, ______

   Rule: ______________________________

8. 80, 60, 70, 50, 60, ______, ______, ______, ______

   Rule: ______________________________

9. 72, 78, 74, 70, 76, ______, ______, ______

   Rule: ______________________________

10. 4, 12, 36, 108, ______, ______, ______, ______

    Rule: ______________________________

Challenge

Have you ever noticed any use of sequences in the real world? Make a list of different situations in which sequences are used.

Going to the Movies
去看电影

With annual ticket（年票）sales of over 7.5 billion（十亿）dollars, going to the movies is a favorite activity for all ages.

Many people attribute the first movie to Eadweard Muybridge. In 1877, Mr. Muybridge took seven pictures of a horse running. He projected（投射）the pictures onto a screen, using a lantern（提灯）. By quickly changing the images, it appeared as if the horse were running. The first full-length feature film（长篇故事片）was made in 1913 by Cecil B. DeMille. Until then, movies were short. Early movies were silent（无声的）and were in black and white. It was not until the late 1920s that sound was added. The development of a new camera（摄影机）in the early 1930s allowed movies to be filmed in color.

Movie technology has advanced a lot since then. Most movies are now being made digitally（数字式）. Computers are used for special effects（特技效果）, animation（卡通片绘制）, editing（剪辑）, and sound design（音效设计）. This technology allows filmmakers（制片人）to easily change color, lighting, and sound to improve the overall effect of a movie. In fact, some animated movies（动画电影）, such as *Toy Story*（《玩具总动员》）, are made completely on the computer.

A movie is said to be a blockbuster（大片）if it is a popular movie that makes a lot of money. Some high-grossing（高票房）blockbusters include *Titanic*（《泰坦尼克号》）($1.8 billion), *Lord of the Rings: The Return of the Ring*（《魔戒：王者归来》）($1.13 billion), and *Pirates of the Caribbean: Dead Man's Chest*（《加勒比海盗：聚魂棺》）($1.1 billion).

The first movie-theater experience began in 1895 when people watched a movie in the basement（地下室）of a café. Today, many megaplex movie theaters（巨型影院）have multiple screens（多屏幕）. Surround sound（环绕立体声）and stadium seating（阶梯座位）complete the moviegoing（观影）experience.

For many people, a trip to the movie theater would not be complete without a stop at the concession stand（小卖部）. Candy（糖果）, hot dogs, soda, and pretzels（椒盐味饼干）are some of the food items moviegoers can purchase to snack（快餐）on while viewing movies. But the most popular item at the concession stand is popcorn（爆米花）. Popcorn was first sold in movie theaters in 1930. Today, it is the best-selling item at concession stands. In 1955, popcorn sales at movie theaters totaled $355 million. Today, movie theaters in the United States sell $15 billion in popcorn each year. That figure accounts for about 30 percent of all popcorn sold in the United States.

Going to the Movies *(cont.)*

What Is the Problem?

Ticket Prices	
Adult ticket	$9.50
Child ticket (ages 2–12)	$6.50
Seniors (65+) (老年票)	$7.50
Concession Stand Menu	
Small popcorn	$4.50
Medium popcorn	$5.25
Large popcorn	$6.00
Small soda	$3.75
Medium soda	$4.00
Large soda	$4.25

You are going to the movie theater to see the latest blockbuster movie. You will have to determine how much the evening at the movies will cost.

Use the information below to answer the following questions. Before you begin solving each problem, think about what information is given and what information you will need to figure out in order to complete the problem. Think about how you will organize the information in order to look for a pattern.

Remember: If you have trouble planning ways to solve each problem, reread it and rethink your strategy. When you have a solution, think about whether your answer makes sense.

Problem-Solving Strategy: Looking for a Pattern

Problem A

The movie theater also offers extra-large popcorn. If the theater follows the pattern in raising the prices for bigger sizes of popcorn, how much would it cost you for extra-large popcorn?

Problem B

Having extra-large popcorn will make you extra thirsty (渴). You would like to request an extra-large soda, as well. If the theater follows the pattern in raising the prices for bigger sizes of soda, how much would it cost you for an extra-large popcorn, an extra-large soda, and the cost of a child's ticket? If you pay $20.00, how much change will you receive?

Problem C

Every year, the theater raises ticket prices by 25 cents and concession prices by 10 cents. In three years, how much will it cost to go to a movie with three of your friends if everyone orders medium popcorn and medium soda?

Hint: Include the cost for tickets and concession. Remember that you plus three of your friends equals four people total. Also, in three years, evaluate the price of the ticket based on age. (Will you be more than 12 years old and need an adult ticket in three years?)

Extension

Plan a class trip to the movies. Create a table that shows how much it would cost for tickets for the entire class to see a movie. If everyone in the class got a medium size popcorn and drink, what would the total cost at the concession stand be? How much more money would each student need if everyone wanted a large popcorn and drink?

Standardized Test Preparation 12

标准考前测试 12

1 Which geometric shape（几何图形）will be created by using reflective symmetry（反射对称）to complete the shape?

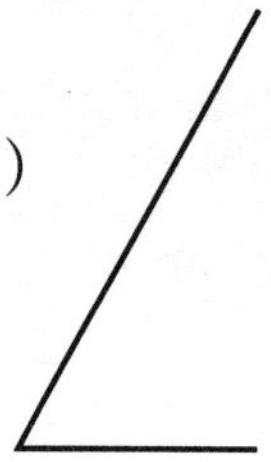

Ⓐ pentagon（五角形）

Ⓑ parallelogram（平行四边形）

Ⓒ triangle（三角形）

Ⓓ square（正方形）

2 If a square has sides that are 4 cm long, what is its perimeter?

Ⓕ 4 cm

Ⓖ 8 cm

Ⓗ 12 cm

Ⓙ 16 cm

3 Which number is half of 1.2 or double 0.3?

Ⓐ 6

Ⓑ 0.06

Ⓒ 0.6

Ⓓ 2.4

4 Choose the best category（范畴）for this statement: *You will ride a camel*（骆驼）*during lunchtime*（午餐时间）*.*

Ⓕ Unlikely

Ⓖ Likely

Ⓗ Certain

Ⓙ Impossible

5 Which geometric figure is a polyhedron（多面体）with eight faces?

Ⓐ octahedron（八面体）

Ⓑ tetrahedron（四面体）

Ⓒ cube（立方体）

Ⓓ cone（圆锥体）

6 Pick one question from this test. Explain how and why you chose your answer.

____

____

____

____

____

____

Lesson 13

Exploring Algebraic Expressions and Equations

探索代数表达式和方程式

Directions: Write the equation（方程式）for each problem. Solve.

1. Rhonda gave her friends clues（暗示）about a mystery（神秘）number she was thinking of. She told them that when her number is tripled（增至三倍）and added to 5, the result is 95. What is the number Rhonda was thinking of?

2. The length of a rectangular sandbox（矩形沙箱）is 5 feet. The area of the sandbox is 40 feet squared. What is the width of the sandbox?

3. Luke gave his friends clues about a mystery number he was thinking of. He told them that when his number is halved（减半）and added to 7, the result is 22. What is the number Luke was thinking of?

Directions: Write the expression described in each problem. Do not solve.

4. There are cakes and pies（派）for sale at the Main Street Bakery（面包店）. There are 10 more cakes for sale than pies. Let *c* represent the number of cakes for sale. Write the expression that represents how many pies are for sale at the Main Street Bakery.

5. At the grocery store（杂货店）, your mom buys 9 more apples than oranges. Let *o* represent the number of oranges. Write the expression that represents the number of apples your mom bought at the store.

6. You have 8 more shirts（衬衣）than pairs of pants（裤子）in your closet（壁橱）. Let *p* represent the number of pairs of pants you have. Write the expression that represents the number of shirts there are in your closet.

Looking for a Pattern Problems

找规律 练习

Directions: Solve each problem and show your work.

Problem A

Write the rule for each number line.

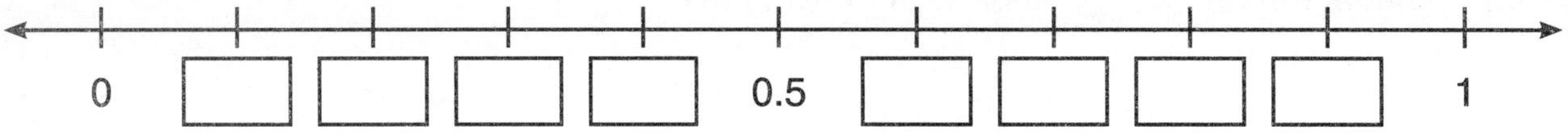

Rule: ______________________________

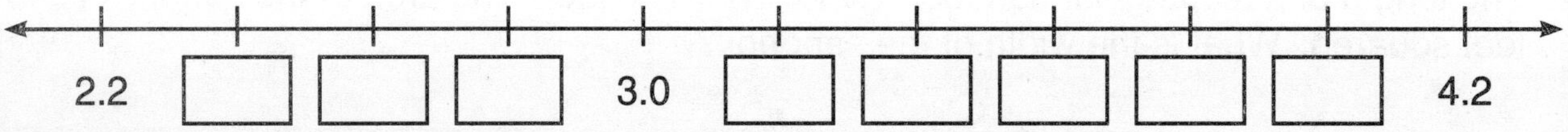

Rule: ______________________________

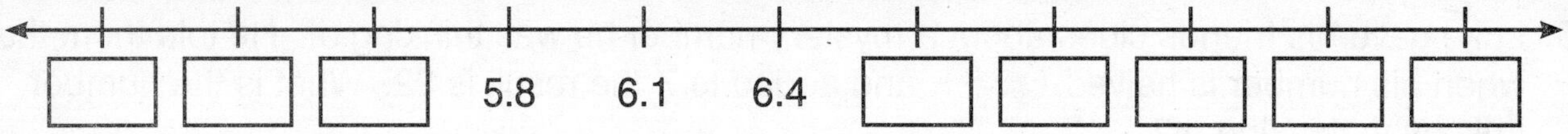

Rule: ______________________________

Problem B

Skip count（跳跃计数） by six.

On the first hundred chart, start with the number six. [Six will be the first number shaded（画上阴影）.] On the second board, start with the number one. (One will be the first number shaded.)Describe and compare the two patterns.

1	2	3	4	5	6	7	8	9	10
11	12	13	14	15	16	17	18	19	20
21	22	23	24	25	26	27	28	29	30
31	32	33	34	35	36	37	38	39	40
41	42	43	44	45	46	47	48	49	50
51	52	53	54	55	56	57	58	59	60
61	62	63	64	65	66	67	68	69	70
71	72	73	74	75	76	77	78	79	80
81	82	83	84	85	86	87	88	89	90
91	92	93	94	95	96	97	98	99	100

1	2	3	4	5	6	7	8	9	10
11	12	13	14	15	16	17	18	19	20
21	22	23	24	25	26	27	28	29	30
31	32	33	34	35	36	37	38	39	40
41	42	43	44	45	46	47	48	49	50
51	52	53	54	55	56	57	58	59	60
61	62	63	64	65	66	67	68	69	70
71	72	73	74	75	76	77	78	79	80
81	82	83	84	85	86	87	88	89	90
91	92	93	94	95	96	97	98	99	100

Standardized Test Preparation 13

标准考前测试 13

1 Which of the following statements is likely?

Ⓐ You will be captured（抓住）by aliens today.

Ⓑ You will win（赢）a billion dollars（十亿美金）today.

Ⓒ You will see a flower somewhere today.

Ⓓ You will become a professional（专业的）football player（足球运动员）this afternoon.

2 What is the volume of a rectangular prism（棱镜）with a length of 6 feet, a height of 2 feet, and a width of 12 feet?

Ⓕ 144 feet cubed

Ⓖ 44 feet squared

Ⓗ 20 feet cubed

Ⓙ 20 feet squared

3 What word best describes these two figures?

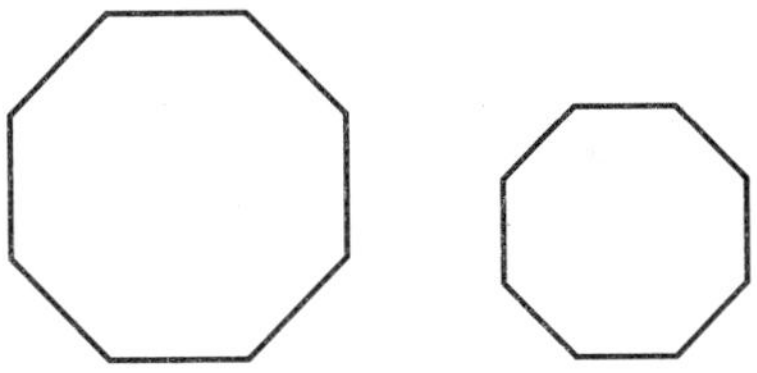

Ⓐ similar（相似）

Ⓑ congruent（全等）

Ⓒ right（直角）

Ⓓ polyhedron（多面体）

4 Which letter is located at coordinates (5, 5)?

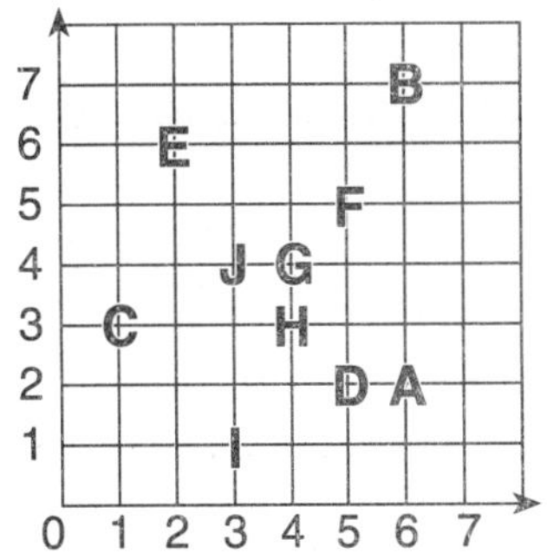

Ⓕ E　　Ⓗ G

Ⓖ D　　Ⓙ F

5 Which state（国家）has the largest area?

California—163,696 sq. mi.
（加利福尼亚）（平方英里）

Alaska—663,267 sq. mi.
（阿拉斯加州）

Texas—268,581 sq. mi.
（德克萨斯州）

Montana—147,042 sq. mi.
（蒙大纳）

Ⓐ California　　Ⓑ Montana

Ⓒ Texas　　Ⓓ Alaska

6 Pick one question from this test. Explain how and why you chose your answer.

____

____

____

____

____

____

Variables and Equations

变量和方程

Directions: Solve each problem using the information given in the problem.

1. Eva and her twin（双胞胎）sister Anna are buying thank you cards（感谢卡）to use for the birthday presents（生日礼物）they received. They created a table to figure out how many boxes of cards to buy. What is the rule for the table? Use the rule to write an expression for this problem and find the number of cards in 4 boxes. Let *b* equal the number of boxes. They need to write 30 thank-you cards. Will they have enough cards if they buy 4 boxes?

Number of boxes (*b*)	1	2	3	4
Number of note cards	8	16	24	

2. There is a big sale（大甩卖）at Mia's favorite clothing store. The table below lists the original price（原价）of each item and the sale price. What is the rule for the table? Use the rule to write an expression for this problem. Let *p* stand for the original price of the item. What is the sale price of an item that is regularly $29.99?

Regular price (*p*)	$9.99	$15.99	$19.99	$29.99
Sale price	$6.99	$12.99	$16.99	

3. Jun is keeping track（记录）of all of the places he visits on family vacations（家庭旅行）. He wants to see how many miles his family drives on each trip. Let *n* represent the number of miles one way to each vacation spot（度假胜地）. Write the expression that tells how Jun can calculate the total number of miles round trip his family drives on each vacation. Next, complete the table using the expression.

Vacation spot	Beach（海滩）	Mountains（山脉）	Desert（沙漠）	Lake（湖泊）	River（河流）
Miles one way (*n*)	78	267	153	241	152
Total miles					

Variables and Equations *(cont.)*

4. There are 12 students in Omar's art class（美术课）. How many students will be in the class if *s* more students enroll（入学登记）? What is the expression that shows the total number of people in Omar's art class? Complete the table showing the total number of students in Omar's class for each given value of *s*.

Number of students enrolling (*s*)	Total number of students
4	
6	
9	
10	
12	

5. There is a sale at Reading Rocks Book Store. The table below lists the original price of each book and the sale price. What is the rule for the table? Use the rule to write an expression for this problem. Let *b* stand for the original price of the item. What is the sale price of a book that is regularly $15.50?

Regular price (*b*)	$3.50	$8.50	$12.00	$14.00	$15.50
Sale price	$2.75	$7.75	$11.25	$13.25	

6. Write your own problem in the space below.

Celebrating a Mexican Holiday
欢度墨西哥节日

The words *Cinco de Mayo*（五月五日节）are Spanish（西班牙语）for May 5. Cinco de Mayo is a popular Mexican national holiday. Can you guess on what day it is celebrated each year? Many people think Cinco de Mayo is Mexico's Independence Day（墨西哥独立日）, but it is not. Mexico celebrates its independence from Spain on September 16.

Cinco de Mayo was first celebrated in Puebla（普埃布拉）, Mexico. Puebla is about 70 miles from Mexico City, which is the capital of Mexico. Mexico City is about 2,200 miles away from New York City.

Cinco de Mayo celebrates the defeat（战败）of a French army（法国陆军）in Mexico in 1862. Over 6,000 French soldiers attacked a Mexican army sheltered in two forts（堡垒）in Puebla. The Mexican army had only about 4,500 soldiers. The Mexicans won the Battle of Puebla. So, Cinco de Mayo celebrates liberty and freedom.

Cinco de Mayo is a symbol（象征）of Mexican unity（团结统一）and patriotism（爱国主义）and is celebrated with parades（游行）, mariachi music（墨西哥流浪乐队）, food, and the customs（风俗）of Mexico. Communities and families that celebrate the holiday may hire mariachi musicians. A mariachi band usually has violins（小提琴）and trumpets（小号）. It also has guitars（吉他）and, sometimes, a Mexican folk harp（墨西哥民间竖琴）.

Mariachi Band Costs

Number of Members	Per Hour Cost of Band
4	$450
5	$500
6	$550
7	$600

Celebrations are not limited to Mexico. People of Mexican heritage（后裔）in the United States celebrate the day, too.

Celebrating a Mexican Holiday *(cont.)*

What Is the Problem?

You are planning the musical entertainment（休闲娱乐）for a Cinco de Mayo party. You must figure out the cost of hiring（雇佣）a mariachi band. You need to work out all of the details for the celebration.

Use the information below to answer the following questions. Before you begin solving each problem, think about what information is given and what information you will need to figure out in order to complete the problem.

Use the information in the problems below to create a table, find the pattern, and answer the questions. Think about how you will organize the information in order to look for a pattern.

Remember: If you have trouble planning ways to solve each problem, reread it and rethink your strategy. When you have a solution, think about whether your answer makes sense.

Problem-Solving Strategy: Looking for a Pattern

Problem A

If the pattern from the table continues as the number of members in the mariachi band increases, how much would it cost to hire a mariachi band with 8 members for 1 hour? How much would it cost to hire a mariachi band with 4 members for 2 hours? What is the difference in price between an 8-member band for 1 hour and a 4-member band for 2 hours?

Problem B

How much would it cost to hire a mariachi band with 8 members for 3 hours? How much would it cost to hire a mariachi band with 4 members for 3 hours?

Problem C

You want to spend exactly $900 on a band for the party. What size band can you hire, and for how long can they play? Are there other options? Create a table that includes the new data. Determine the new range and median for the Mariachi Band Costs table.

Extension

Make a line graph to show the trend（趋势）in cost. Predict the cost of a 15-member mariachi band.

Standardized Test Preparation 14

标准考前测试 14

1 Which number is divisible by 1, 2, 3, and 9?

Ⓐ 13

Ⓑ 29

Ⓒ 36

Ⓓ 27

2 What is the best estimate for the length of a jump rope（跳绳游戏）?

Ⓕ 6 inches

Ⓖ 6 yards

Ⓗ 6 feet

Ⓙ 6 meters

3 What are the coordinates for point D?

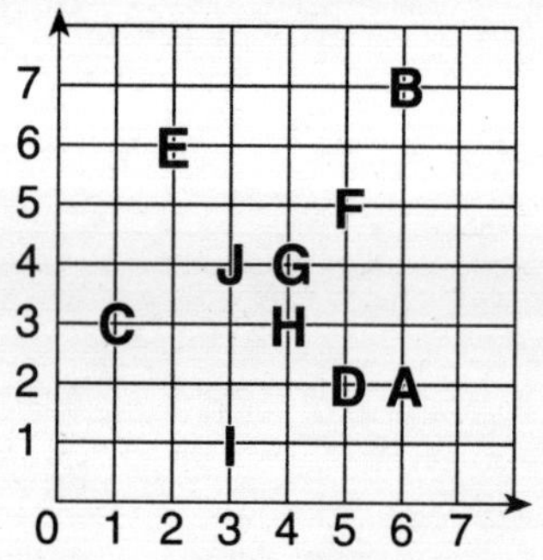

Ⓐ (2, 5) Ⓒ (5, 2)

Ⓑ (4, 1) Ⓓ (0, 5)

4 What is the perimeter of the fence（围墙）?

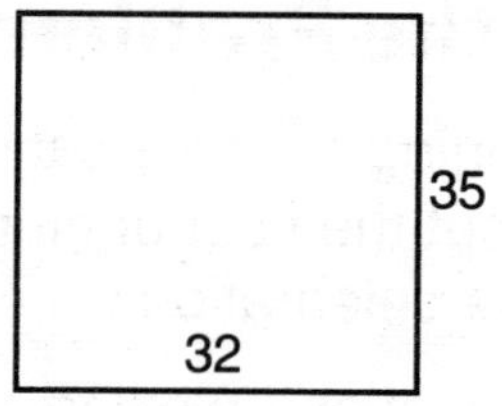

Ⓕ 35 + 32 + 35 + 32

Ⓖ 35 x 32 + 35 x 32

Ⓗ 35 ÷ 32 + 35 ÷32

Ⓙ 35 – 32 x 35 – 32

5 Which number shows the numbers in order from least to greatest?

Ⓐ 9,556 9,565 9,655 9,491

Ⓑ 9,565 9,655 9,491 9,556

Ⓒ 9,556 9,655 9,565 9,491

Ⓓ 9,491 9,556 9,565 9,655

6 Pick one question from this test. Explain how and why you chose your answer.

____

____

____

____

____

____

Lesson 15

How Long Does It Take?

需要多久呢？

Directions: Read and solve each of the following problems involving（涉及的）time.

1. Calculate the length of time in hours and minutes between each of the following sets of times:

a. From 7:20 a.m. to 12:00 noon ____________

b. From 11:11 to 19:22 ____________

c. From 6:17 to 18:04 ____________

d. From 3:10 p.m. to 2:05 a.m. ____________

e. From 14:12 to 6:00 ____________

f. From 7:11 a.m. to 11:07 p.m. ____________

g. From 12:00 to 15:00 ____________

h. From 10:10 a.m. to 3:01 p.m. ____________

i. From 1:00 to 23:17 ____________

j. From the start of school to first recess ____________

k. From the end of lunch to the end of school ____________

2. A movie at the theater starts at 7:10 p.m. and finishes at 9:25 p.m. How long does it run?

____

3. Lisa and Brian went to see a play, which lasted 2 hours and 12 minutes. The play ended at 2:04 p.m. What time did it start?

____

4. A bicyclist（骑脚踏车者）left home at 6:30 a.m. She rode until 8:10 a.m., stopped for a 20-minute break（休息）, and then cycled home, which took 1 hour and 50 minutes. What time was it when she got back home?

____

5. Paul and Vickie drove from Chicago（芝加哥）, Illinois（伊利诺斯州）, to Cincinnati（辛辛那提）, Ohio（俄亥俄州）. It took 3 hours and 50 minutes to drive from Chicago to Indianapolis, and it took 2 hours 25 minutes to drive from Indianapolis（印第安纳波利斯）to Cincinnati.

a. How many minutes did the trip take in all? ____________________

b. How many hours and minutes did it take? ____________________

Looking for a Pattern Group Problems

找规律 分组练习

Directions: Solve Problem A and show your work. Using the strategy you've studied this week, write and solve your own problems for Problem B and Problem C.

Problem A

1. Write the fraction below each circle.
2. Draw the next three fraction circles in the pattern.
3. Write the fraction below each circle.

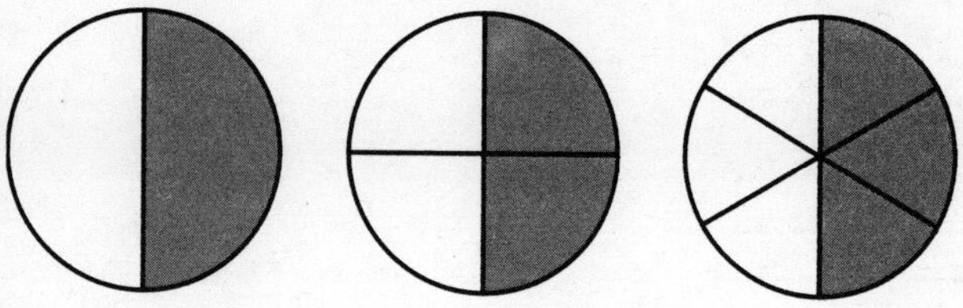

________ ________ ________ ________ ________ ________

Problem B: ____________________

Show work:

Answer: ____________________

Problem C: ____________________

Show work:

Answer: ____________________

Standardized Test Preparation 15

标准考前测试 15

1 Irma received these grades on her math tests: 80, 70, 65, 80, 60, 60, 75, 75, 80, 85. What is the mode?

Ⓐ 80

Ⓑ 75

Ⓒ 60

Ⓓ 65

2 Which line is perpendicular（垂直于）to AB?

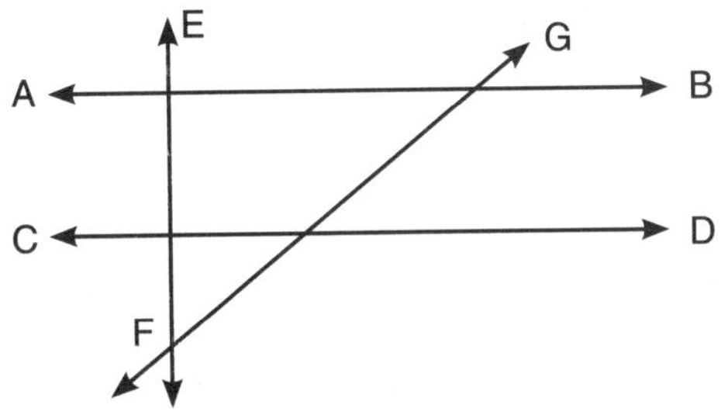

Ⓕ $\overleftrightarrow{EF}$　　Ⓗ $\overleftrightarrow{FG}$

Ⓖ $\overleftrightarrow{CD}$　　Ⓙ None

3 What is the volume of the shape?

Ⓐ 9 units squared

Ⓑ 8 units cubed

Ⓒ 8 units squared

Ⓓ 9 units cubed

4 Which of these letters does not illustrate（说明） reflective symmetry?

Ⓕ **V**

Ⓖ **R**

Ⓗ **T**

Ⓙ **U**

5 Maria has a box shaped like a hexagon（六角形）. What is the perimeter of this box if each side has a measurement of 14 in.?

Ⓐ 12 in.

Ⓑ 24 in.

Ⓒ 48 in.

Ⓓ 84 in.

6 Pick one question from this test. Explain how and why you chose your answer.

____

____

____

____

____

____

How Much? How Far? How Long?

有多少？有多远？有多长？

Directions: Read each question carefully. Think about what you need to find out, decide what units need to be used, and then solve each problem.

1. Each member of Brian's family tried to guess the weight of Brian's birthday cake.

Mom	$5\frac{1}{2}$ lb.	Dad	3 lb. 10 oz.
Alan	2 lb.	Jackie	$3\frac{1}{4}$ lb.
Uncle（叔叔）Ed	4 lb. 3 oz.	Aunt（阿姨）Becky	$1\frac{1}{2}$ lb.

The cake weighed 60 ounces.

a. Who had the closest guess? ________________

b. Whose guess was the next closest? __________

c. Which guess was furthest off? ____________

d. Insert the **less than** (<) or **greater than** (>) symbols in each circle to compare the guesses.

3 lb.	◯	$5\frac{1}{2}$ lb.	$1\frac{1}{2}$ lb.	◯	2 lb.
3 lb.10 oz.	◯	2 lb.	$3\frac{1}{4}$ lb.	◯	3 lb.10 oz.
$3\frac{1}{4}$ lb.	◯	4 lb. 3 oz.			

2. Anthony's mother wanted to knit（织）him a sweater（毛线衣）. The pattern said that she needed 12 balls of yarn. If each ball of yarn is 115 yards long, how much yarn will be used to make the sweater?

a. yards? ____________ **b.** feet? ____________ **c.** inches? ____________

3. Lisa's mother bought 9 yards of fabric（布）to make a new dress. To make the dress, she needed 4 feet.

a. How much fabric will she have left after cutting out the dress? ____________

b. Does she have enough fabric to make two dresses? Explain your answer. __________

____

4. A bottle of bubble bath（泡沫浴）holds（容纳）1 gallon.

a. If 32 ounces are used each time, how many bubble baths does each bottle make?

____

b. If a family of four each took a bath every day, how many days will they be able to use the bubble bath? Explain your answer. ____________________

____

Drawing a Diagram

绘制图表

Drawing a diagram or picture of a word problem often helps you visualize（形象化）, or picture in your mind, the information in a problem more clearly.

You can draw a diagram using simple symbols or pictures. Drawing diagrams also helps you keep track of the information in problems that have more than one step. Diagrams allow you to organize your thoughts, which makes the problems simpler.

Using Drawings in Math

Simple drawings help you picture the information in a problem in your mind.

- You should not spend a lot of time on the drawing. A "quick draw" will help you record what is happening so that you can solve the problem.
- Sometimes you can draw a picture of the actual object in the problem. Other times, it works better to use a drawing to represent an object in the problem. For example, a square may represent an animal.

Problem: Fence Posts（栅栏柱）

The Problem

You have been asked to find out how many fence posts would be needed if a post was placed every 2 yards along a 10-yard section of wood fence.

Understanding the Problem

- *What do we know?*

  In the fence, a post will be placed every 2 yards. The total section of fence will be 10 yards.

- *What do we need to find out?*

  How many posts you need.

Planning and Communicating a Solution

By drawing a picture, it is easy to figure out how many posts are needed to build the fence.

Drawing a picture helps you solve the problem because you can picture the solution, so you will make fewer mistakes.

For this picture, you do not need to actually draw fence posts. You can draw small lines to represent the posts.

You might mentally calculate 10 divided by 2 equals 5, so 5 posts are needed. However, if you make a drawing to show the fence and posts, you will see that this equation does not give you the correct answer. You need one for the starting point and another at the end of the fence!

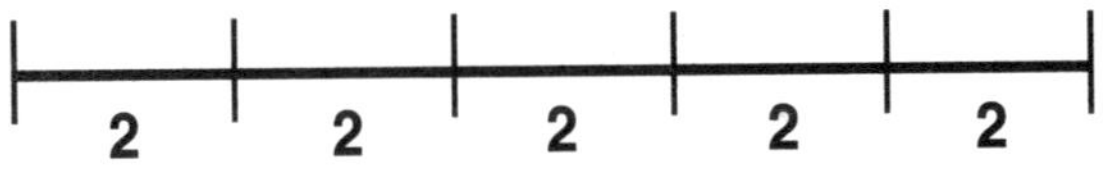

- *Do you see the answer?*

  You need 6 posts, each 2 yards apart.

Reflecting and Generalizing

Always ask, "Can I solve this problem with simple math calculations, or should I draw a diagram to better understand what is happening in the problem?" In this problem, without the picture you might have gotten confused about the total number of fence posts. The picture helps you visualize the scene in the problem.

Extension

If you had 14 fence posts that were placed 2 yards apart, how long would the total fence be?

Standardized Test Preparation 16
标准考前测试 16

1 What is the quotient?

$9.5 \div 10 = \_\_\_$

Ⓐ 95

Ⓑ 0.95

Ⓒ 950

Ⓓ 0.095

2 Which of the following is the most likely weight of a book?

Ⓕ 20 oz.

Ⓖ 20 lb.

Ⓗ 10 lb.

Ⓙ 2 oz.

3 Simplify $\frac{6}{16}$ by finding the greatest common factor.

Ⓐ $\frac{3}{8}$

Ⓑ $\frac{3}{6}$

Ⓒ $\frac{3}{4}$

Ⓓ $\frac{5}{8}$

4 Triangles ABC and DEF are congruent. Which of the following is true?

?

Ⓕ Their perimeters are unequal, but the areas are equal.

Ⓖ The areas are unequal, but the perimeters are equal.

Ⓗ The perimeters and areas are both unequal.

Ⓙ The perimeters and areas are both equal.

5 Which of the following statements is certain?

Ⓐ Dinosaurs（恐龙） eat meat.

Ⓑ Everyone in your community eats meat.

Ⓒ Carnivores（食肉动物） eat meat.

Ⓓ Adults eat meat.

6 Pick one question from this test. Explain how and why you chose your answer.

____

____

____

____

____

____

Lesson 17

Tiles
拼方块

Directions: There are four ways of arranging four square tiles so that each tile meets along at least one edge. Write the perimeter of each arrangement below it (the side of a tile = 1 inch).

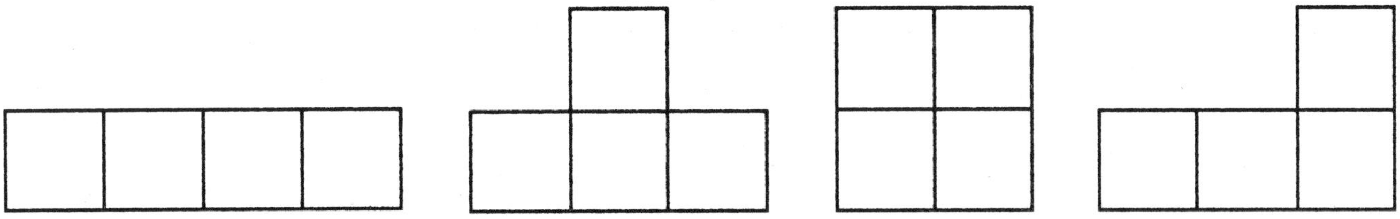

1. P = ________ **2.** P = ________ **3.** P = ________ **4.** P = ________

5. Which arrangement above has the smallest perimeter? Why?

____

____

6. How many different ways can you arrange five tiles? Use the squares below. Write the perimeter of each arrangement next to it.

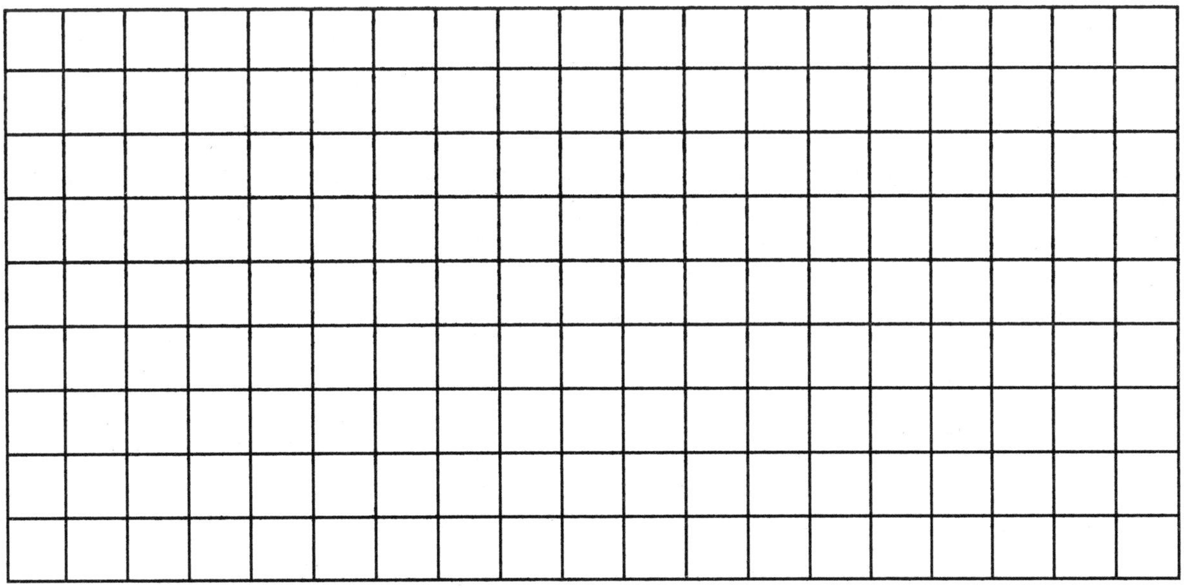

Challenge

Directions: On one-inch grid（方格）paper, draw arrangements of tiles with a perimeter of exactly 12 inches. Can you use 6, 7, 8, and 9 squares?

Tiles *(cont.)*

Directions: Use this page to complete the Challenge problem from page 67. Draw arrangements of tiles with a perimeter of exactly 12 inches. Can you use 6, 7, 8, and 9 squares?

The Summer Olympics

夏季奥运会

The Summer Olympic Games are held every four years. The Olympics last for about two weeks. In 2004, they were held in Athens（雅典）, Greece（希腊）. Over 200 nations competed in the games. Each nation sent its best athletes（运动员）to compete in the Games. About 300 events（项目）were held in 2004. These included events in swimming, gymnastics（体操）, and track and field（田径赛）.

The Olympics have many swimming events. The swimming pool is in the shape of a rectangle. The pool is 50 meters long. It must be at least 21 meters wide. The pool is divided into lanes for the racing events. In these events, the swimmers swim in their own lanes against other swimmers. The fastest swimmer wins the race! In some races, the swimmers swim on their stomachs. In other races, they swim on their backs.

Other water sports, such as water polo（水球）and diving（跳水）, are also Olympic events（奥运会项目）. These events take place in different pools than the swimming events. When a city is chosen to host（主办）the Olympic games, it must build many new pools just for the Olympic swimming events.

The Summer Olympics *(cont.)*

What Is the Problem?

The Olympic Games promote（发扬）athletic competition（体育竞争）at its highest level. The Summer Olympics have been popular throughout their existence.

Diagrams can help you visualize the given information. You can use diagrams to solve the problems below about Olympic-size pools（游泳池）.

Use the information on page 69 and below to answer the questions. Before you begin solving the problems, be sure to locate the information you will need.

Remember: If you have trouble planning ways to solve each problem, reread it and rethink your strategy. When you have a solution, think about whether or not your answer makes sense.

Problem-Solving Strategy: Drawing a Diagram

Problem A

Draw a diagram of an Olympic-size pool that is 25 meters wide. Tell how much longer the pool is than it is wide.

Problem B

Draw a diagram of an Olympic-size pool that has 10 lanes（泳道）. Each lane is 3 meters wide. What is the width of the pool? Label the length and width of the pool.

Problem C

Water polo must be played in a pool that is no longer than 30 meters. The pool cannot be wider than 20 meters. How do the pools for Olympic swimming and water polo compare? Draw diagrams to show each pool. Could water polo be played in the Olympic swimming pool? Tell why or why not.

Hint: Make sure your drawings show the correct shape for each pool.

Extension

If your school wanted to hold a mini-olympics, where would you hold your competitions? Measure the length and width of the place where you will hold your olympics. Draw and label a diagram of the location. Divide the diagram into four parts. As a class decide which events will be held in each section. Do you have enough space to hold your mini-olympics?

Standardized Test Preparation 17

标准考前测试 17

1 Patrick rode the Ferris Wheel（摩天轮）for \$3.75 and the Loopyloop for \$2.95. How much less was the Loopyloop ride?

- Ⓐ \$1.80
- Ⓑ \$0.90
- Ⓒ \$0.70
- Ⓓ \$0.80

2 What was the average（平均）monthly temperature for June shown on the line graph?

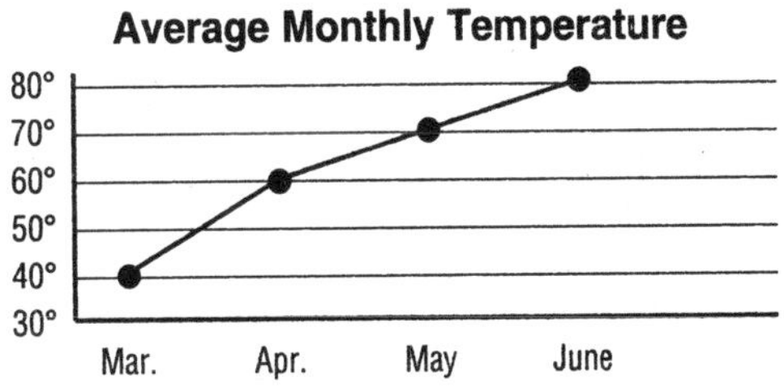

- Ⓕ 60°
- Ⓗ 80°
- Ⓖ 70°
- Ⓙ 50°

3 What is the best estimate for the length of a crayon?

- Ⓐ 3 inches
- Ⓑ 3 centimeters
- Ⓒ 3 millimeters
- Ⓓ 3 feet

4 What is 3.5 doubled?

- Ⓕ 6.5
- Ⓖ 1.75
- Ⓗ 70
- Ⓙ 7

5 Which line segment is neither perpendicular nor parallel to any other line segment?

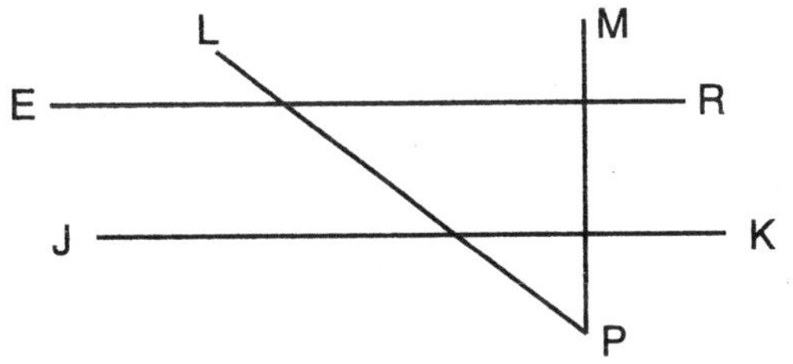

- Ⓐ $\overline{ER}$
- Ⓒ $\overline{PL}$
- Ⓑ $\overline{JK}$
- Ⓓ $\overline{PM}$

6 Pick one question from this test. Explain how and why you chose your answer.

____

____

____

____

____

____

Area According to Units
根据单位计算面积

Directions: Use the measurements next to each rectangle below to calculate the area of the rectangle. Record your answer inside each rectangle. [**Note:** These shapes are not drawn to scale（比例）.]

1. 7 mm, 9 mm ____________

2. 12.5 in., 6 in. ____________

3. 120 yd., 5 yd. ____________

4. 1.6 m, 2.5 m ____________

5. 86 cm, 31 cm ____________

6. 15 ft., 28 ft. ____________

7. 30 in., 19 in. ____________

8. 71 mm, 71 mm ____________

9. 17.1 m, 71 m ____________

10. 77 mi.（英里）, 21 mi. ____________

Drawing a Diagram Problems

绘制图表 练习

Directions: Solve each problem and show your work.

Problem A

Rico has 7 classes a day. The 7 classes are Art, English, Math, Music, P.E., Reading, and Science. What time does each class start?

1. English is before break.
2. Math is after lunch.
3. Reading is the last class of the day.
4. P.E. is the second class of the day.
5. Art is after 9 a.m. but before break.
6. Music is right after the break.

Daily Schedule（日程表）

8:00–8:45	__________
8:45–9:30	__________
9:30–10:15	__________
10:15–10:30	Break
10:30–11:15	__________
11:15–12:00	__________
12:00–12:45	Lunch
12:45–1:30	__________
1:30–2:15	__________

Problem B

Michelle wants to build a pen in her backyard（后院） for her potbellied（大腹便便的） pig, Porker. She wants each side to be 10 feet long. Every 2 feet, Michelle wants to put a fence post. How many fence posts does Michelle need?

Standardized Test Preparation 18
标准考前测试 18

1 Which of these decimals（小数） has the least value?

Ⓐ 0.194

Ⓑ 0.2

Ⓒ 0.1847

Ⓓ 0.21

2 Jean traveled $\frac{20}{45}$ of the bike route in a morning ride. Express the distance as a simplified fraction.

Ⓕ $\frac{4}{9}$

Ⓖ $\frac{4}{13}$

Ⓗ $\frac{5}{9}$

Ⓙ $\frac{2}{9}$

3 What is the perimeter of this rectangle?

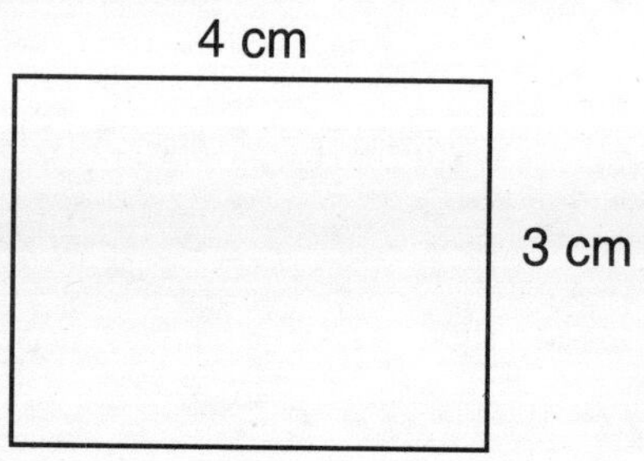

Ⓐ 14 cm

Ⓑ 12 cm

Ⓒ 11 cm

Ⓓ 7 cm

4 Choose the best category for this statement: *You will blink*（眨）*your eyes some time in the next 10 minutes.*

Ⓕ Impossible

Ⓖ Unlikely

Ⓗ Likely

Ⓙ Certain

5 Find the range in this set of numbers.

(14, 27, 26, 16, 24, 25, 23, 28, 24, 26, 14, 24)

Ⓐ 12

Ⓑ 10

Ⓒ 14

Ⓓ 9

6 Pick one question from this test. Explain how and why you chose your answer.

____

____

____

____

____

____

Lesson 19

Fill It Up

填一填

Directions: Count（数） the number of cubic units（立方体） in each figure. Remember to count the cubic units you cannot see. Then write the volume of each figure in cubic units.

1.

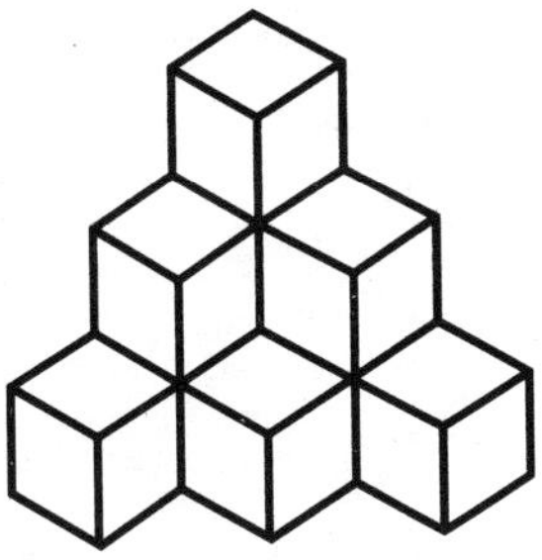

$V =$ ________

2.

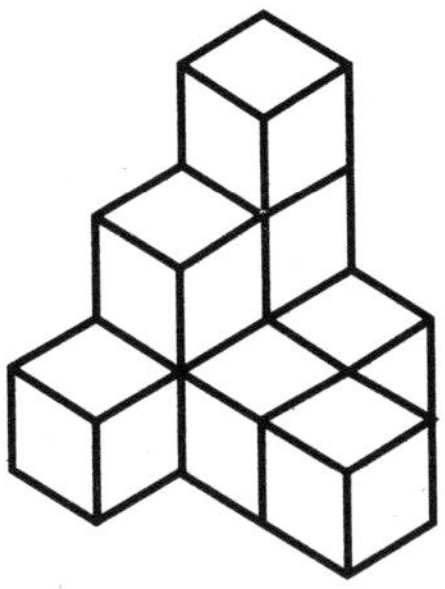

$V =$ ________

3.

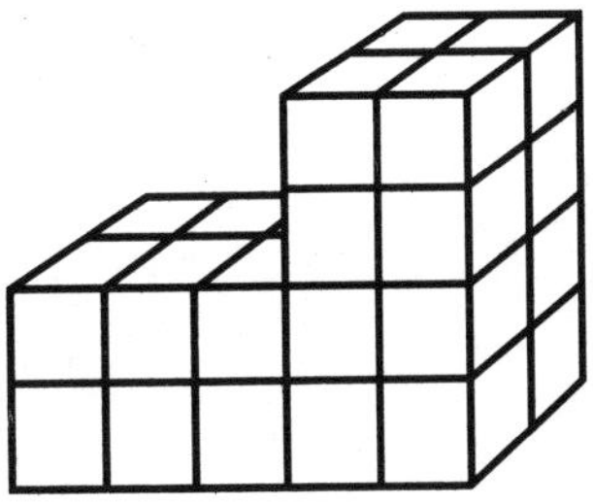

$V =$ ________

4.

$V =$ ________

Directions: For problems 5–10, round your answer to the nearest whole number（整数）.

5. Find the volume.

$l = 11.3$ mm

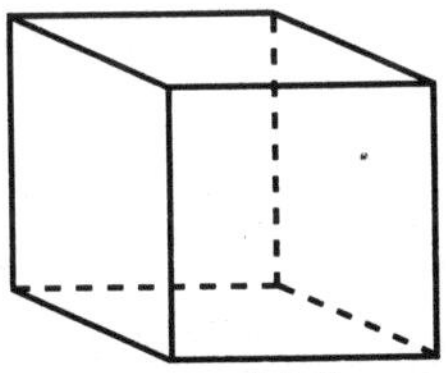

$h = 11.3$ mm

$w = 11.3$ mm

$V =$ ________

6. Find the volume.

$l = 6.2$ mm

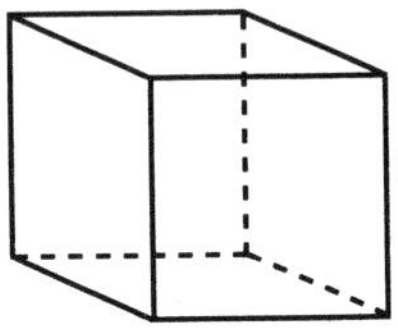

$h = 6.2$ mm

$w = 6.2$ mm

$V =$ ________

7. Find the height of the rectangular prism.

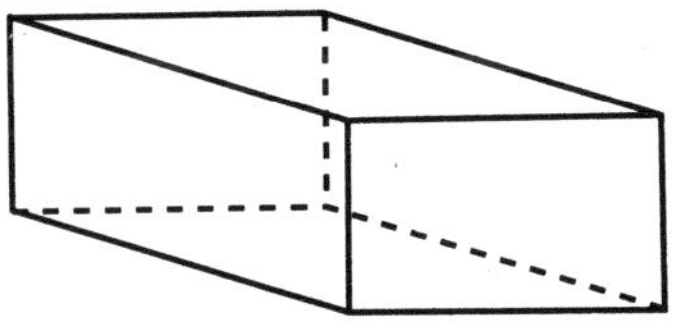

$V = 108$ in.^{3}

$l = 9$ in.

$w = 4$ in.

$h =$ ________

8. Find the volume.

$V =$ ________

$h = 11.3$ mm

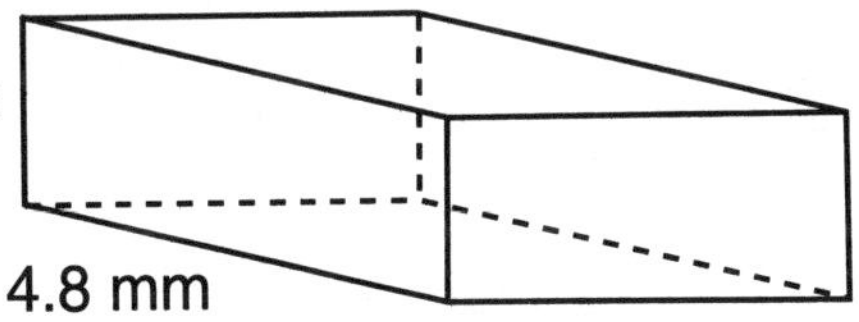

$l = 4.8$ mm

$w = 13$ mm

9. Find the length of the side of this cube.

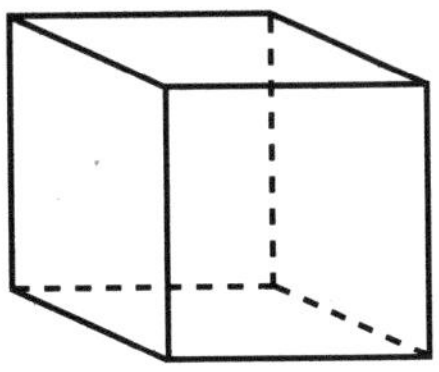

$V = 512$ ft.^{3}

$s =$ ________

10. Find the height.

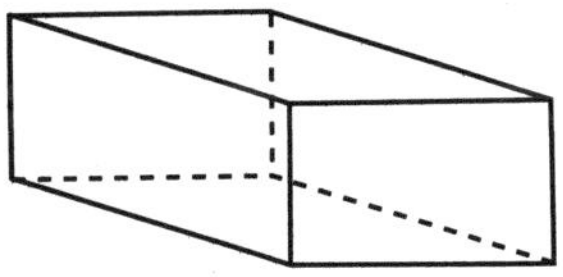

$V = 96$ in.^{3}

$l = 8$ in.

$w = 3$ in.

$h =$ ________

The Tallest Mountains in the United States

美国最高的山

In some parts of the United States, there are tall mountains. All of the tallest mountains are in the West. In fact, Alaska（阿拉斯加州）boasts 16 of the highest peaks（最高峰）in the United States. The tallest is Denali（德纳里峰）. It is 20,320 feet (6,190 m) high! It is covered in snow and ice. People first climbed to the top of this mountain in 1913.

Denali is also called Mount McKinley（麦金利山）after President William McKinley. *Denali* means "the great one."

The second tallest mountain in the United States is Mount St. Elias（圣伊莱亚斯山）. It is 18,008 feet (5,490 m) tall. Mount Foraker is the third tallest mountain. It is 17,400 feet (5,300 m) tall.

Many people climb these mountains every year, but it is very difficult. They must climb them with the help of guides. Climbers of these tall mountains must bring a lot of equipment and carry their own food and supplies to the tops of the mountains.

The Tallest Mountains in the United States *(cont.)*

What Is the Problem?

Many people climb mountains. It is exciting（令人兴奋的）and fun, but it can be dangerous（危险）. Imagine that you are a mountain climber. Your guide（导游）is leading you up Mount Foraker（福克拉山）. You climb up the mountain for 6 days. Then you rest for 2 days. You climb again 1 day and then rest for 1 day. The next day, you climb to the top of the mountain. You stay there for 3 days. Then you take 3 days to climb down the mountain.

Use the information on page 76 and above to draw diagrams and answer the questions. Before you begin solving the problems, be sure to locate the information you will need.

Remember: If you have trouble planning ways to solve each problem, reread it and rethink your strategy. When you have a solution, think about whether or not your answer makes sense.

Problem-Solving Strategy: Drawing a Diagram

Problem A

Draw a diagram to show how long your entire trip will take. How many days will the trip take?

Hint: You might draw tally marks for each day of the trip.

Problem B

Draw a diagram to find out how many days it will take you to get to the top of the mountain.

Hint: Don't count the days that you rest at the top.

Problem C

Draw a line to plot（标绘）the route up the mountain. Read through the problem, sentence by sentence, and plot the travel for each day. Use rising lines like this **/** to mark each day spent climbing. Use falling lines like this **** to mark each day spent going downhill（下山）. Use flat（平坦的）lines like this — to show each day spent resting. How many days were spent resting?

Extension

As a class, research the 10 tallest mountains in the world. Draw a diagram showing the height of each of the mountains. Did the mountains discussed in the article make the top 10 list? If so, what were their ranks? If not, how much taller would they have to be to make the list?

Standardized Test Preparation 19
标准考前测试 19

1 Which answer lists all the factors of 16?

Ⓐ 1, 2, 8, 16

Ⓑ 1, 2, 4, 8

Ⓒ 1, 2, 4, 8, 10

Ⓓ 1, 2, 4, 8, 16

4 Which three-dimensional figure has only square faces?

Ⓕ rectangular prism

Ⓖ triangular prism

Ⓗ square-based pyramid

Ⓙ cube

2 What is the next number in this sequence?

1, 3, 6, 10, 15, ___

Ⓕ 20

Ⓖ 19

Ⓗ 21

Ⓙ 28

5 What was the highest daily temperature for Tuesday shown on the line graph?

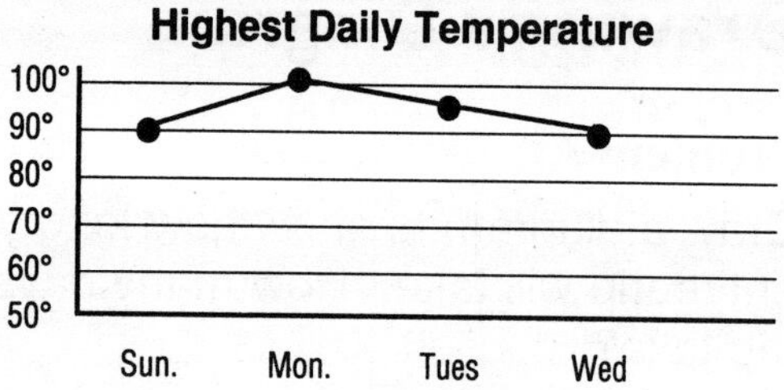

Ⓐ 90° Ⓒ 80°

Ⓑ 95° Ⓓ 85°

3 What is the area of this envelope (信封)?

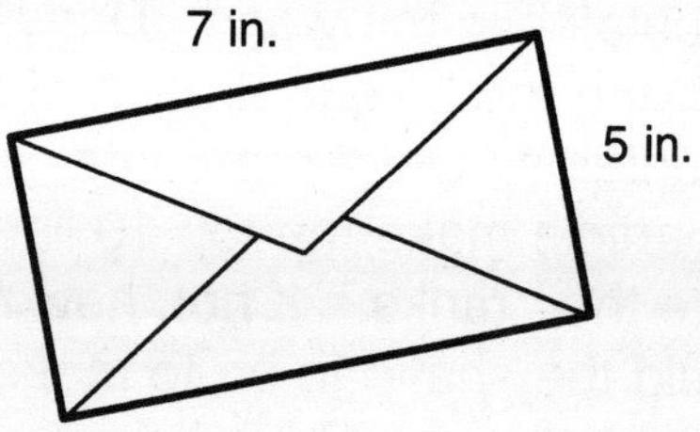

Ⓐ 12 in.^{2} Ⓒ 55 in.^{2}

Ⓑ 35 in.^{2} Ⓓ 70 in.

6 Pick one question from this test. Explain how and why you chose your answer.

____

____

____

____

____

____

Name Coordinates
姓名坐标

Directions: Plot the letters of your first name and your last name anywhere on the coordinate grid. Example: Guanlin zhou (7 letters, 4 letters).

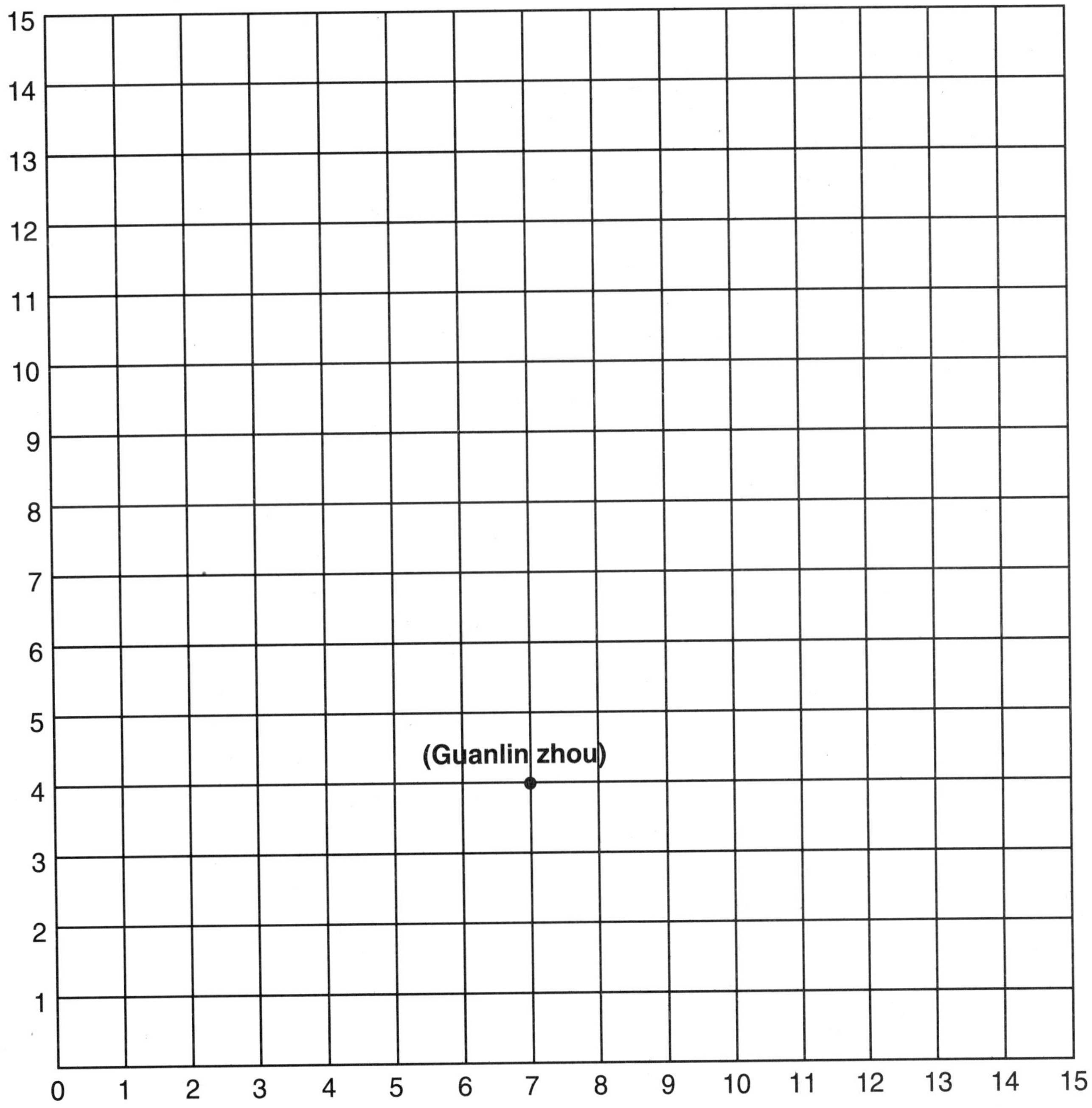

Drawing a Diagram Group Problems

绘制图表 分组练习

Directions: Solve Problem A and show your work. Using the strategy you've studied this week, write and solve your own problems for Problem B and Problem C.

Problem A

Arrange 12 trees in a circle. The trees are numbered 1 to 12. The first tree is an apple tree. Every third tree after that is also an apple tree.

1. Out of the remaining trees, every fifth one is an orange tree.
2. Out of the remaining trees, every third one is a peach tree.
3. The rest of the trees are cherry trees.

How many of each kind of tree are there?

Problem B: ____________________

____

____

____

____

____

Show work:

Answer: ____________________

Problem C: ____________________

____

____

____

____

____

Show work:

Answer: ____________________

Standardized Test Preparation 20

标准考前测试 20

1 What is the area of the sheet of paper?

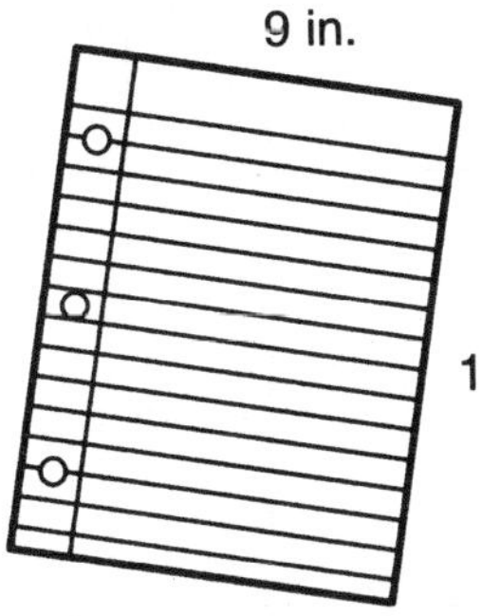

Ⓐ 18 in.^{2}

Ⓑ 22 in.^{2}

Ⓒ 40 in.^{2}

Ⓓ 99 in.^{2}

2 What is the volume of the cylinder?

V*= π *r*^{2}*h

Ⓕ 251.2 inches cubed

Ⓖ 502.4 inches cubed

Ⓗ 251.2 inches squared

Ⓙ 502.4 inches squared

3 Mike recorded these heights in inches for 10 of his classmates: 60, 45, 52, 54, 53, 60, 57, 56, 57, 60. Which is the mode?

Ⓐ 60

Ⓑ 57

Ⓒ 56

Ⓓ 45

4 What are the coordinates for point E?

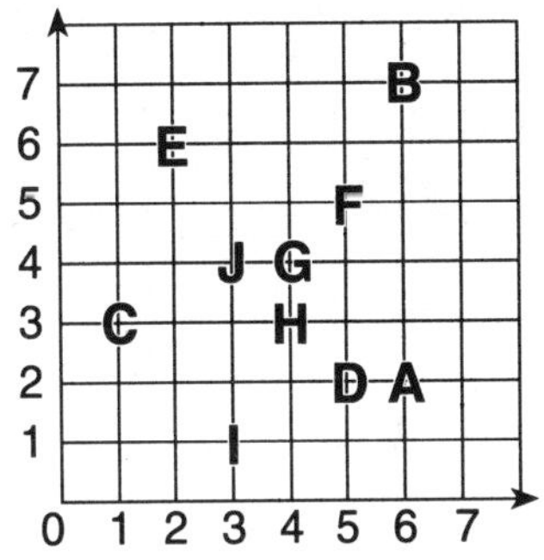

Ⓕ (6, 2) Ⓗ (2, 6)

Ⓖ (2, 5) Ⓙ (5, 5)

5 Which letter of the alphabet illustrates perpendicular lines?

Ⓐ **J**

Ⓑ **S**

Ⓒ **T**

Ⓓ **K**

6 Pick one question from this test. Explain how and why you chose your answer.

____

____

____

____

____

____

Lesson 21

Parallel and Perpendicular

平行和垂直

Directions: Plot one capital letter showing parallel lines and one capital letter showing perpendicular lines. Choose one shape that shows parallel lines and plot it on the grid. Record the points. Choose one shape that shows perpendicular lines and plot it on the grid. Record the points.

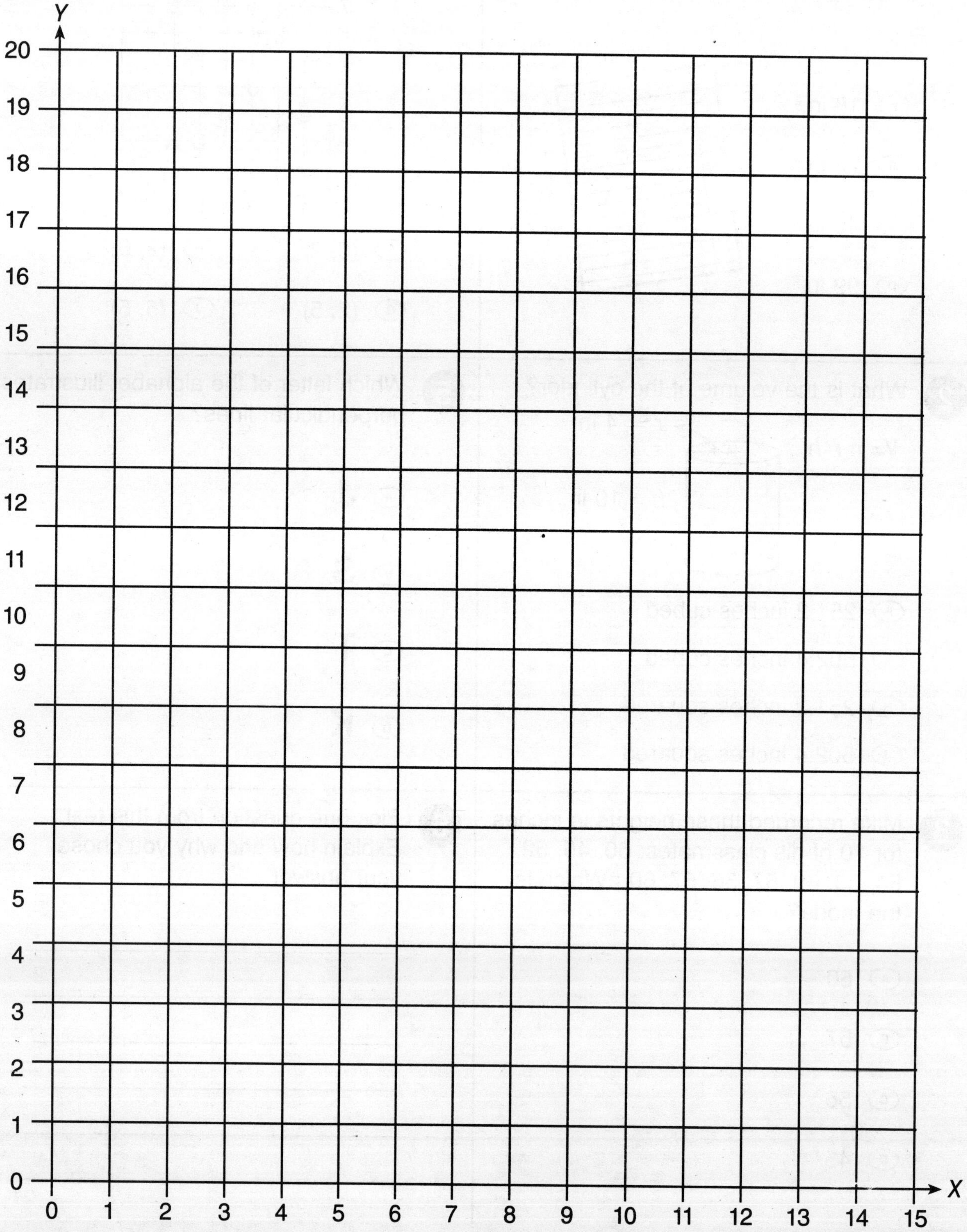

Space to the Right

填写右侧坐标

Directions: Use the graph below to help you plot the coordinates for each problem.

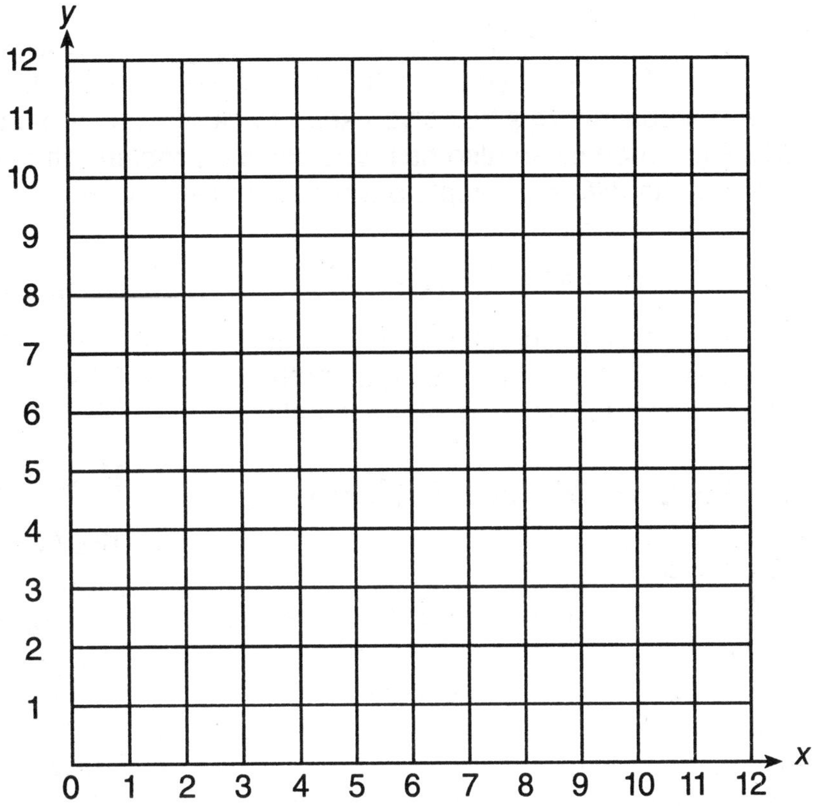

1. Plot each set of coordinates listed below to make rectangles. Then identify each pair of parallel lines by their coordinates.

   a. (2, 3), (6, 3), (6, 9), (2, 9)

   b. (11, 1), (12, 1), (12, 12), (11, 12)

   c. (7, 0), (10, 0), (10, 4), (7, 4)

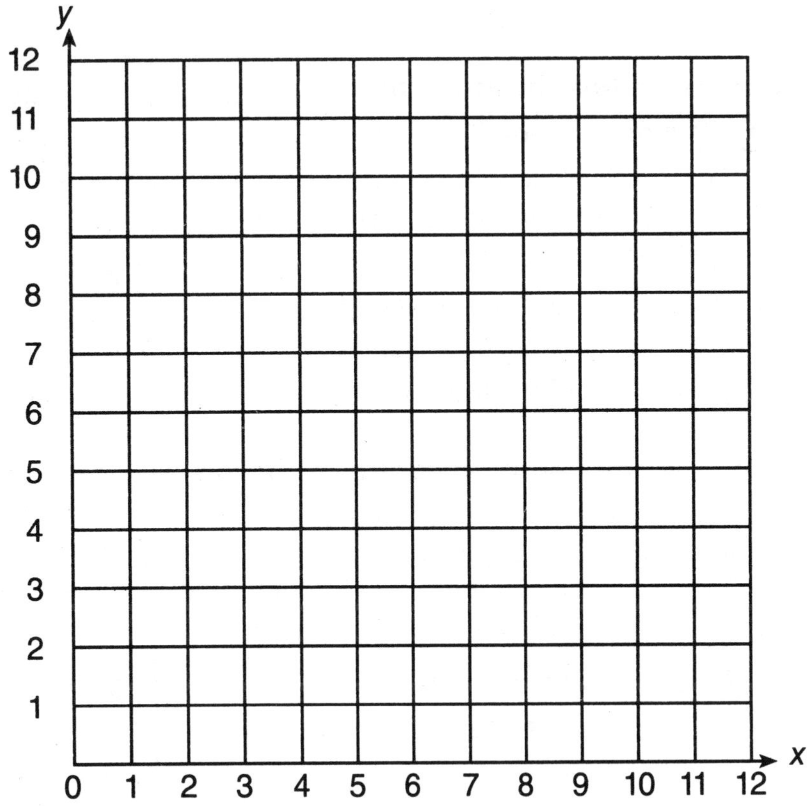

2. Plot each of the perpendicular lines A and B. Then name the coordinates for the point where the lines intersect.

   a. A (1, 2), (6, 2)
   B (5, 5), (5, 2)

   ____

   b. A (7, 1), (12, 1)
   B (9, 1), (9, 3)

   ____

   c. A (2, 4), (2, 10)
   B (2, 8), (4, 8)

   ____

   d. A (11, 11), (11, 4)
   B (7, 7), (11, 7)

   ____

Acting It Out or Using Concrete Materials

演示或用小玩具解题

Using objects, such as counters or blocks, to represent people, places, or things in a problem often helps you find the answer. These objects can be moved through the steps of the problem. This helps you keep track of what is happening as you solve the problem. Making a model can also help you solve a problem. It can also be valuable to act out the roles of the different people or objects in the problem.

Use Concrete Materials

When a problem contains large numbers or gives directions for building something, it may be best to use concrete materials.

Problem: Movie Tickets

The Problem

John is waiting in line to buy tickets at the movies. There are 50 people in front of him, and John is very impatient (不耐烦). Each time a person is served, John cuts in front of 2 people. How many people will be served before John reaches the front of the line?

Understanding the Problem

- *What do we know?*

  There are 50 people in front of John. Each time someone buys tickets, John cuts in front of 2 people.

- *What do we need to find out?*

  How many people will be served before John reaches the front of the line?

Planning and Communicating a Solution

It helps to visualize the movement of people in the problem by actually moving some counters or blocks as John and the other people in line would be moving forward.

- *Do you see the answer?*

  17 people will be served.

Act It Out!

When a problem has things being moved around, acting it out can make it easier to find the solution.

Problem: Pet Steps

The Problem

Fluffy was lying asleep on the middle step. Milo sat 3 steps above Fluffy. There were 2 steps between Milo and the top step. How many steps were there altogether?

Understanding the Problem

- *What do we know?*

  Fluffy is on the middle step. Milo sat 3 steps above Fluffy. There are 2 steps between Milo and the top step.

- *What do we need to find out?*

  How many steps there are altogether.

Planning and Communicating a Solution

Draw a staircase (楼梯). Use a counter to show Fluffy's position on the middle step. Place a second counter 3 steps above Fluffy to represent Milo. Leave 2 steps empty. The third step above Milo should be the top step. Count to find out how many steps there are above Fluffy. Then, add the same number below Fluffy's position.

- *Do you see the answer?*

  There are 13 steps altogether.

Reflecting and Generalizing

Both problems are easier to solve if you use real objects or act out the steps.

Standardized Test Preparation 21

标准考前测试 21

1 Bryan has 350 fliers（传单） to distribute（分发） among 5 classes. How many fliers will each class receive?

Ⓐ 70 fliers

Ⓑ 80 fliers

Ⓒ 74 fliers

Ⓓ 700 fliers

2 Choose the best category for this statement: *Someone from your class will travel to the moon in your lifetime*（一生）.

Ⓕ Impossible

Ⓖ Likely

Ⓗ Certain

Ⓙ Unlikely

3 The number 7,850 is

Ⓐ closer to 7,900 than it is to 7,800.

Ⓑ less than 8,000 but more than 7,900.

Ⓒ greater than 7,800 but less than 7,850.

Ⓓ halfway between 7,800 and 7,900.

4 An airplane（飞机） left Washington, D.C.,（华盛顿特区） at 1:22 p.m. and arrived at Boston（波士顿） at 3:06 p.m. How long was the trip?

Ⓕ 2 hr. 16 min.

Ⓖ 1 hr. 16 min.

Ⓗ 1 hr. 44 min.

Ⓙ 1 hr. 34 min.

5 What is the area of a rectangle that has a length of 7 ft. and a width of 2 ft.?

Ⓐ 7 $ft.^2$

Ⓑ 14 $ft.^2$

Ⓒ 24 $ft.^2$

Ⓓ 42 $ft.^2$

6 Pick one question from this test. Explain how and why you chose your answer.

____

____

____

____

____

____

Shape Configurations

图形组合

Directions: Find and list similar shapes and congruent shapes.

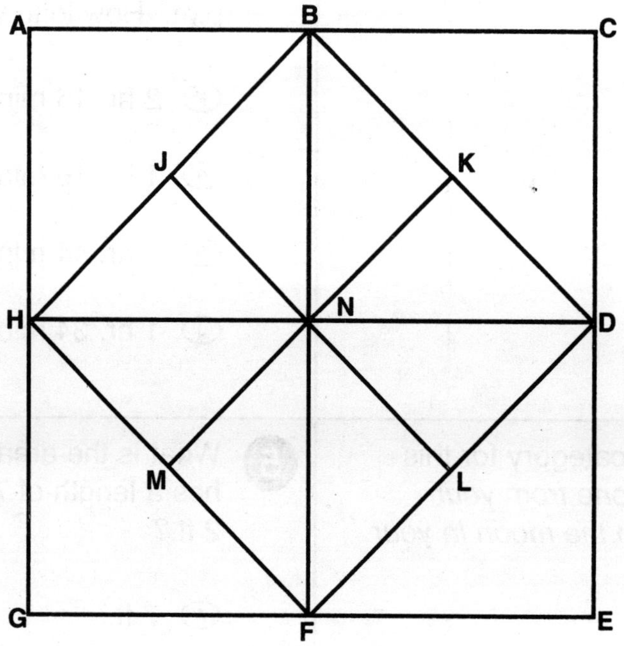

Figure B

Directions: Shade Figure C below according to the directions on the next page.

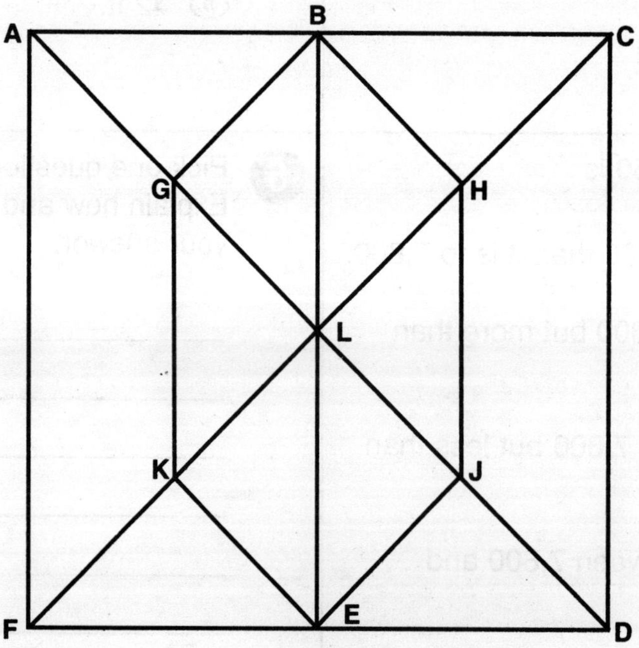

Figure C

Similar and Congruent Shapes
相似图形和全等图形

Section A

Directions: Color Figure C according to the direction in each problem below.

1. Color the shape ABG green.
2. Color the shape BGKL purple.
3. Color the shape CHJD orange.
4. Color the shape KLJE blue.

Directions: List congruent and similar shapes for each part of Figure C.

	Congruent Shapes	**Similar Shapes**
5. Shape ABG		
6. Shape BGKL		
7. Shape CHJD		
8. Shape KLJE		

9. Write the names of three parallel lines:

____

____

____

10. Write the names of three perpendicular line segments:

____

____

____

Similar and Congruent Shapes *(cont.)*

Section B

Directions: Draw a congruent shape and a similar shape for each shape listed below.

		Congruent Shape	Similar Shape
11.	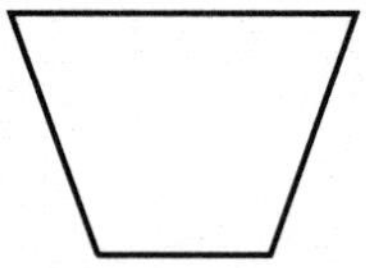	____________	____________
12.	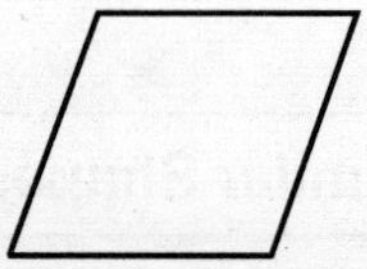	____________	____________
13.	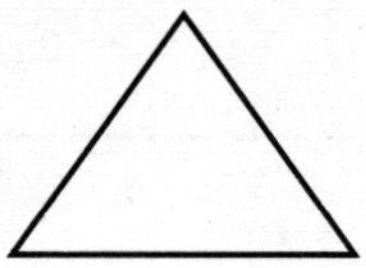	____________	____________
14.	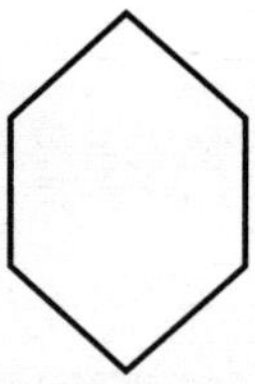	____________	____________

Directions: Color Figure C according to the directions in each problem below.

15. Color the shape BLJHC red. Then, find as many congruent shapes as you can.

Congruent Shapes: ______________________________________

____

____

16. Color the shape AGKEF yellow. Then, find as many congruent shapes as you can.

Congruent Shapes: ______________________________________

____

____

Liquids
液 体

16 ounces（液盎司） **2 cups**（量杯） **1 gallon**（加仑）

Capacity refers to what an item can hold. The volume of a item is how much space can be occupied（占用） inside. Volume can refer to liquid（液体） or solid（固体） material.

At the grocery store, you will find references to volumes of liquids. For example, a milk jug usually comes in a half-gallon jug or a gallon jug. Water bottles usually tell you the liters （升）, pints（品脱）, or fluid ounces（液盎司）. All of these terms refer to volume of liquid items.

While the term *liters* refers to the metric system（公制）, much of the food in the United States is measured in standard volume units—ounces, cups, quarts, pints, or gallons.

2 cups = 1 pint = 16 ounces
4 cups = 1 quart = 32 ounces
16 cups = 1 gallon = 128 ounces

Liquids *(cont.)*

What Is the Problem?

You can use real containers（容器）or drawings of real containers to solve the problems about measuring the volume of liquid.

Use the information from the previous page and in the problems below to answer the questions. Before you begin solving each problem, think about what information is given and what information you will need to figure out in order to complete the problem.

Remember: If you have trouble planning ways to solve each problem, reread it and rethink your strategy. When you have a solution, think about whether your answer makes sense.

Problem-Solving Strategy: Acting It Out or Using Concrete Materials

Problem A

You have a gallon jug of milk and a measuring cup that measures 1 cup. How many cups can you use to empty the gallon of milk?

Problem B

If you have a gallon jug of milk and there are 6 people who want to each drink a pint of milk, how many cups are left?

Problem C

If there are 14 people who want to each drink a cup of milk, how many pints remain in the milk jug?

Extension

Use a milk jug drawing, counters, and the *Gallon Conversions* diagram on page 145 to create at least one word problem of your own. Use Problems A, B, and C as models if you need help.

Standardized Test Preparation 22

标准考前测试 22

1 What is the most likely amount of liquid in a can of soda?

Ⓐ 12 qt.

Ⓑ 12 L

Ⓒ 12 oz.

Ⓓ 12 pt.

2 What are the coordinates for point G?

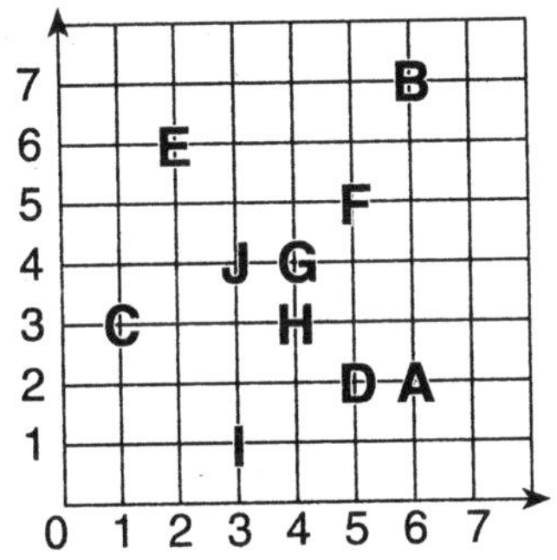

Ⓕ (4, 6)   Ⓗ (6, 2)

Ⓖ (4, 4)   Ⓙ (4, 5)

3 Which state has the smallest area?

Arizona（亚利桑那州）—113,998 sq. mi.

Montana（蒙大纳）—147,042 sq. mi.

Colorado（科罗拉多州）—104,094 sq. mi.

Nevada（内华达州）—110,561 sq. mi.

Ⓐ Arizona   Ⓒ Colorado

Ⓑ Montana   Ⓓ Nevada

4 Traci checked on the amount of change her friends had in their pockets. Determine the range. (55¢, 34¢, 88¢, 14¢, 5¢, 22¢, 98¢, 76¢, 93¢)

Ⓕ 93¢

Ⓖ 88¢

Ⓗ 55¢

Ⓙ 44¢

5 Which could be used to find the perimeter of a rectangle?

Ⓐ 2 + 2 + 4 + 4

Ⓑ 2 x 2 + 4 + 4

Ⓒ 2 x 4

Ⓓ 4 ÷ 2

6 Pick one question from this test. Explain how and why you chose your answer.

____

____

____

____

____

____

Lesson 23

Acting It Out or Using Concrete Materials Problems
演示或用小玩具解题 练习

Directions: Solve each problem and show your work.

Problem A

Write the numbers 36, 37, 38, 39, 40, 41, 42, 43, and 44 in the squares below so that each row and column equals 120.

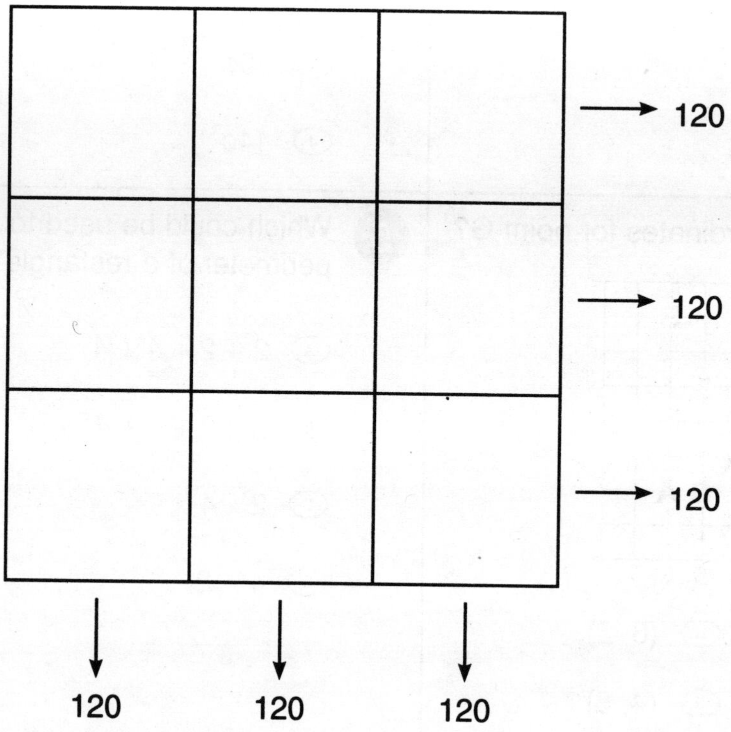

Problem B

Ethan owed（欠）his dad $10.00. He paid his dad $2.00 but then borrowed（借）$1.00. Ethan then paid his dad $7.00 but soon had to borrow $3.00. Ethan borrowed $5.00 more, but later paid his dad $9.00. Has Ethan paid his dad back all of the money he owed him, or does he still owe his dad money?

Grids and Shapes

网格和图形

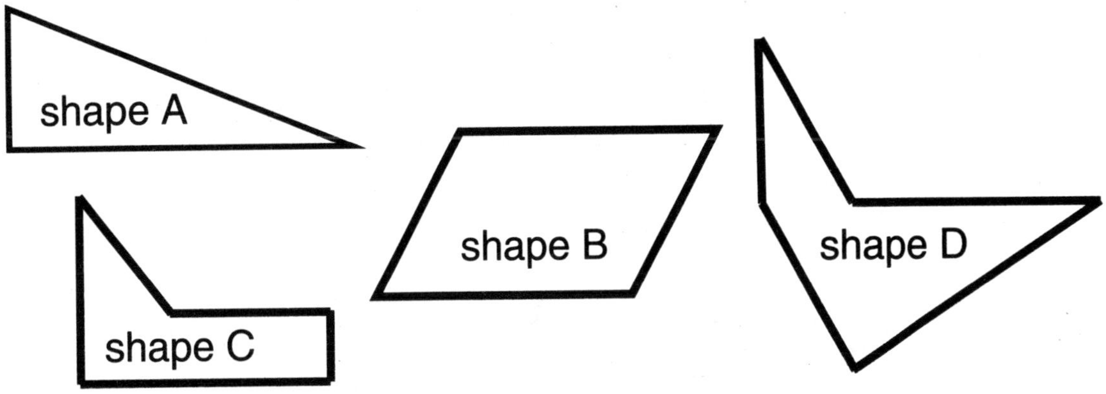

cut here

Directions: Cut out the shapes above. Then, with their rotations, translations or reflections, put they on the coordinate grids.

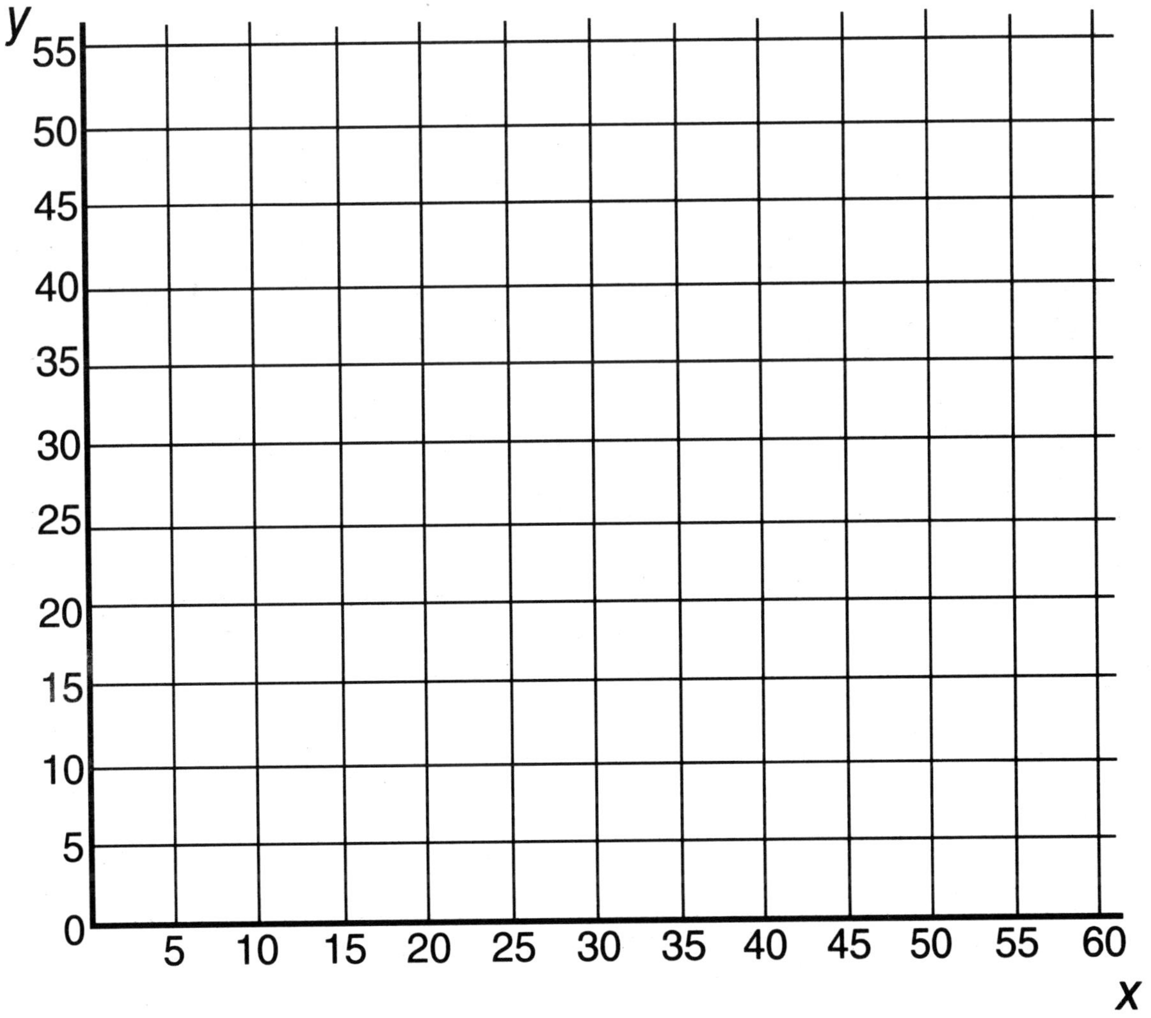

Grids and Shapes *(cont.)*

This section of the page is left blank so you can cut out the manipulatives on the previous page.

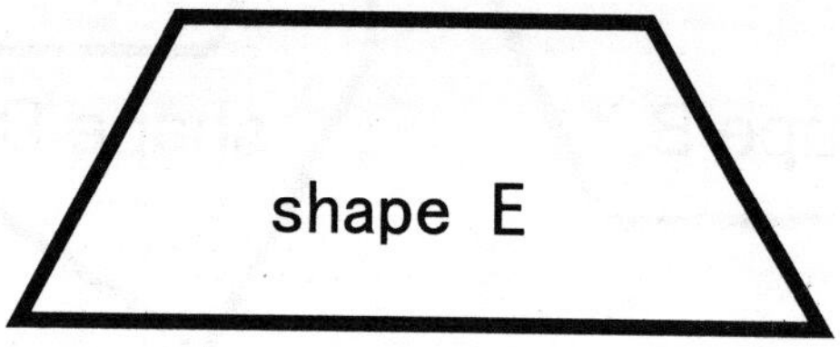

cut here

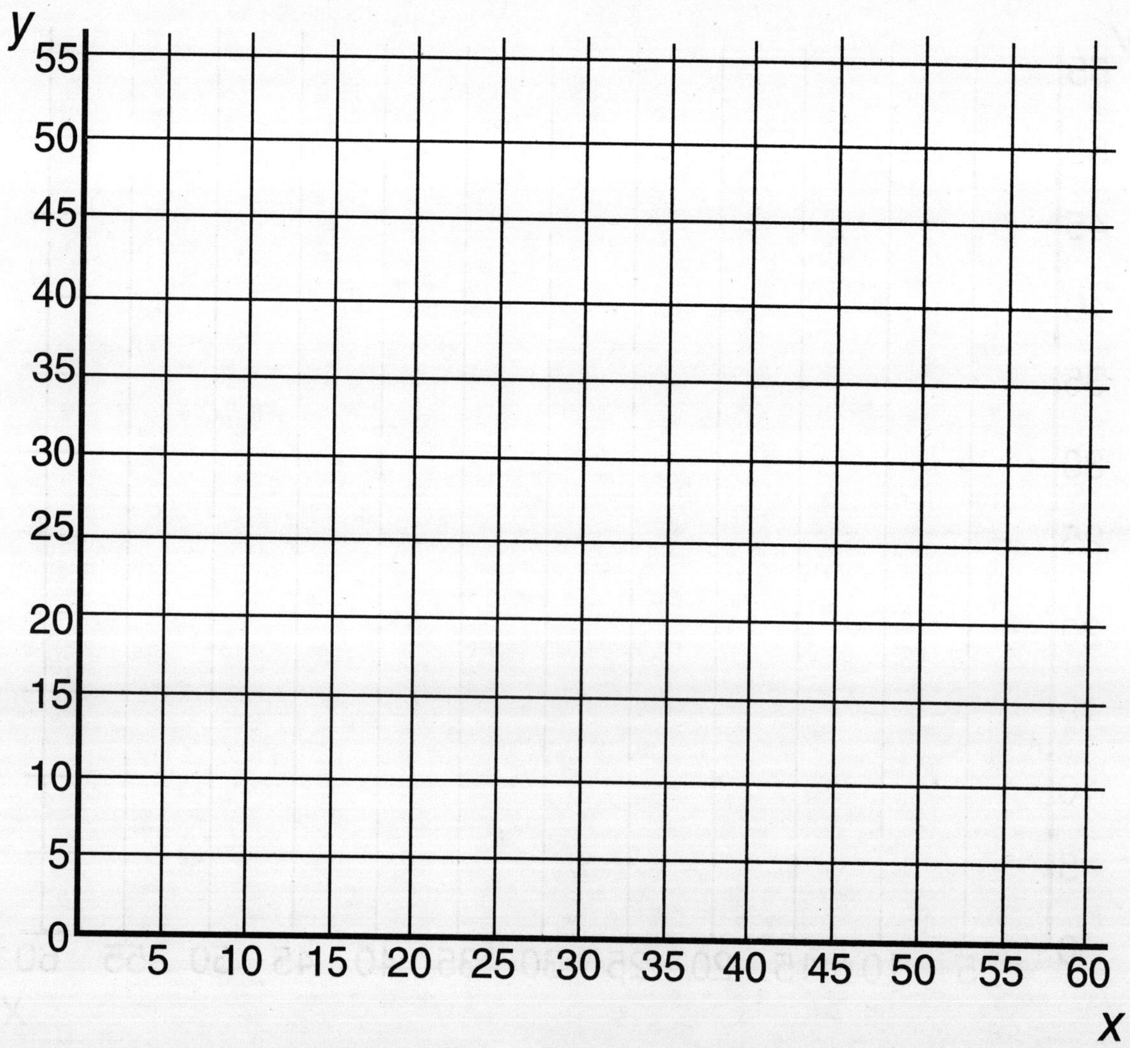

Shape Coordinates Recording Sheet

图形坐标记录单

Section A

Directions: Use the shapes from page 93 to complete the transformation described in each problem. Then, write the coordinates for each corner of each shape.

1. Shape A ____
2. Shape A translation ____
3. Shape A rotation ____
4. Shape A reflection ____
5. Shape B ____
6. Shape B translation ____
7. Shape B rotation ____
8. Shape B reflection ____
9. Shape C ____
10. Shape C translation ____
11. Shape C rotation ____
12. Shape C reflection ____

Shape Coordinates Recording Sheet *(cont.)*

Section B

Directions: Use the shapes from page 93 to complete the transformation described in each problem. Then, write the coordinates for each corner of each shape.

13. Shape D ______________________________

14. Shape D translation ______________________________

15. Shape D rotation ______________________________

16. Shape D reflection ______________________________

17. Shape E ______________________________

18. Shape E translation ______________________________

19. Shape E rotation ______________________________

20. Shape E reflection ______________________________

Standardized Test Preparation 23

标准考前测试 23

1 What is the name of a triangle with two equal sides?

Ⓐ equilateral

Ⓑ isosceles

Ⓒ right

Ⓓ scalene

2 Which of the following has the greatest weight?

Ⓕ 0.9 kg

Ⓖ 0.33 kg

Ⓗ 0.1842 kg

Ⓙ 0.41 kg

3 The actual surface of an object or the space it covers is the ______.

Ⓐ diameter

Ⓑ perimeter

Ⓒ angle

Ⓓ area

4 Javier reads 35 minutes a day for 14 days. How many hours and minutes does this equal?

Ⓕ 8 hr. 10 min.

Ⓖ 9 hr. 10 min.

Ⓗ 7 hr. 50 min.

Ⓙ 8 hr. 20 min.

5 What is the volume of the cylinder（圆柱体）?

$V = \pi r^2 h$

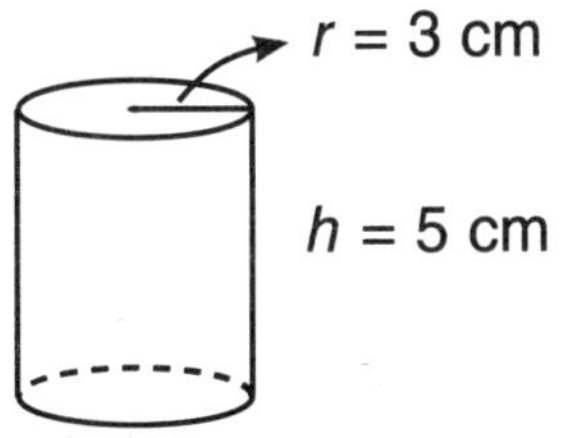

Ⓐ 141.3 centimeters cubed

Ⓑ 94.2 centimeters squared

Ⓒ 141.3 centimeters squared

Ⓓ 94.2 centimeters cubed

6 Pick one question from this test. Explain how and why you chose your answer.

____

____

____

____

____

____

Venn Diagram
维恩图

Directions: Look closely at the diagram. On a separate sheet of paper, write a statement about each quadrilateral. List as much information as you can.

1.

Quadrilaterals (四边形)

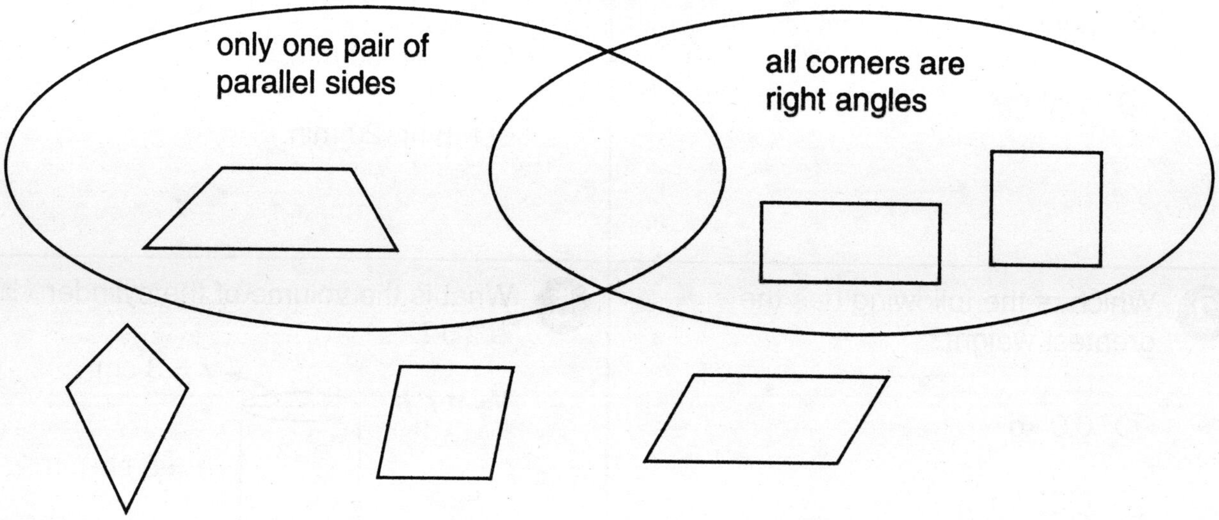

Directions: Draw each of the following shapes in the correct place on the diagram below.

2.

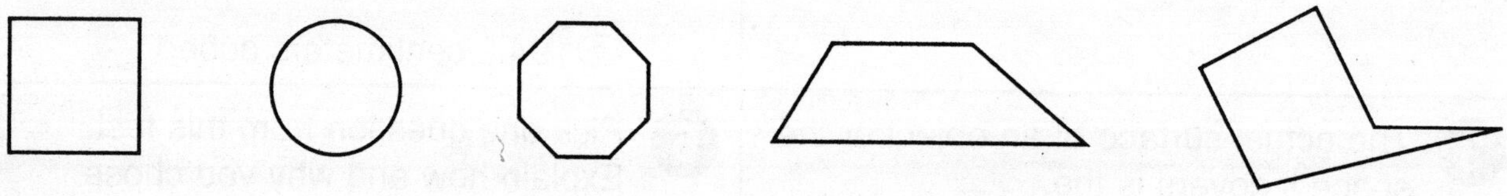

Two-Dimensional Shapes (二维图形)

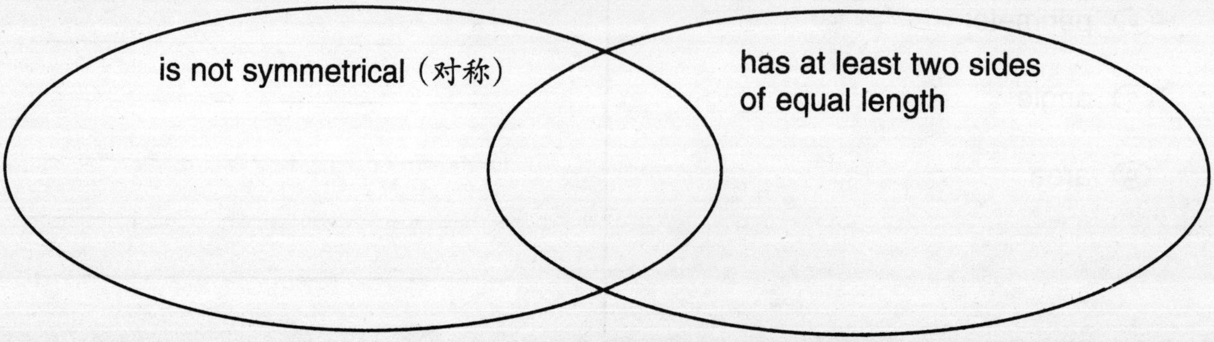

The Box Factory

纸盒工厂

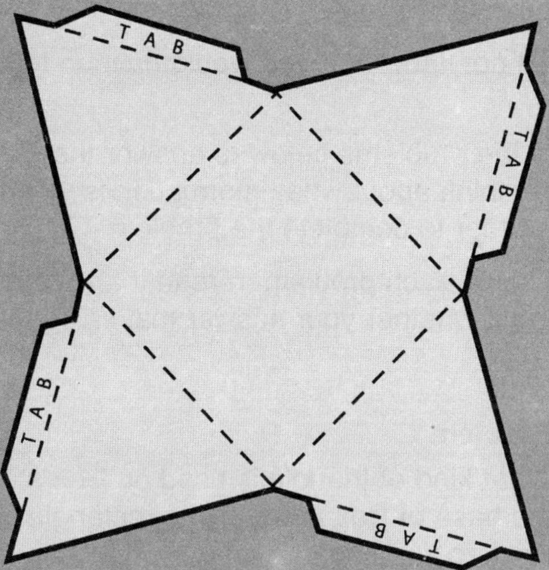

Box A

Have you ever wondered how a cardboard box（纸箱）is made? Boxes are made in a factory. The factory has rolls（卷）of cardboard. Sometimes a machine is used to glue layers（粘接层）of cardboard together. This makes the cardboard stronger. A machine might cut the sheet into an open-box shape. This shape can be folded into a box later. Machines can be set to cut boxes in many different sizes and shapes.

Boxes are used for many different things. Some boxes, such as cereal boxes, hold food. Other boxes hold toys. Some boxes are even made to hold drinks! Juice boxes have a special coating inside them to hold the liquid. No matter what the box holds, it was made in a box factory.

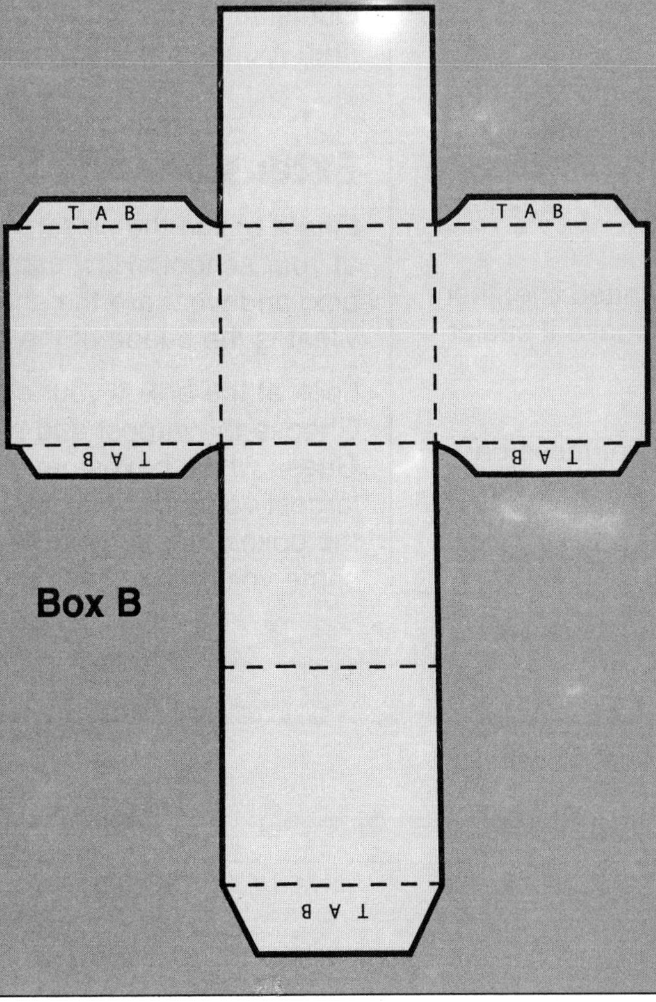

Box B

The Box Factory *(cont.)*

What Is the Problem?

You ordered several boxes from a factory. You did not know that you would have to fold them into boxes.

Use the information from the previous page and in the problems below to answer the questions. Before you begin solving each problem, think about what information is given and what information you will need to figure out in order to complete the problem.

Remember: If you have trouble planning ways to solve each problem, reread it and rethink your strategy. When you have a solution, think about whether your answer makes sense.

Problem-Solving Strategy: Acting It Out or Using Concrete Materials

Problem A

Copy the box design or trace the shapes from page 99 onto a sheet of paper. Carefully cut out the shapes. Fold both boxes on the dotted lines. How many flat faces does each box have?

Hint: First fold along each dotted line; then fold it all together. This will make it easier.

Problem B

What are the shapes of the flat faces on the boxes? What are the names of the completed box shapes?

Problem C

What kind of triangle is used on Box A? Is the base of Box A square or rectangle? How do you know?

One side of Box B is $1\frac{1}{8}$ inch. What is its area rounded to the hundredths place（百分位）?

Extension

Find a box in the classroom or somewhere at your school. How many faces are on the box, and what are the shapes of the faces? What is the shape of the box?

Look at the boxes your classmates found. Choose the largest and smallest boxes. Guess which boxes have the smallest and largest volumes. Calculate the volumes of the boxes. Do any of the boxes have the same volumes but different shapes?

Standardized Test Preparation 24

标准考前测试 24

1 Which of the following has the least value?

Ⓐ 0.243

Ⓑ 0.21

Ⓒ 0.198

Ⓓ 0.3

2 Which number can be divided evenly into 91?

Ⓕ 12

Ⓖ 15

Ⓗ 7

Ⓙ 3

3 Which is the most likely length of a robin, which is a small bird?

Ⓐ 10 in.

Ⓑ 10 ft.

Ⓒ 10 m

Ⓓ 10 mm

4 Which number is smaller than 7,564 but larger than 7,393?

Ⓕ 7,359

Ⓖ 6,564

Ⓗ 7,419

Ⓙ 7,664

5 Which line segment is parallel to line segment $\overline{GR}$?

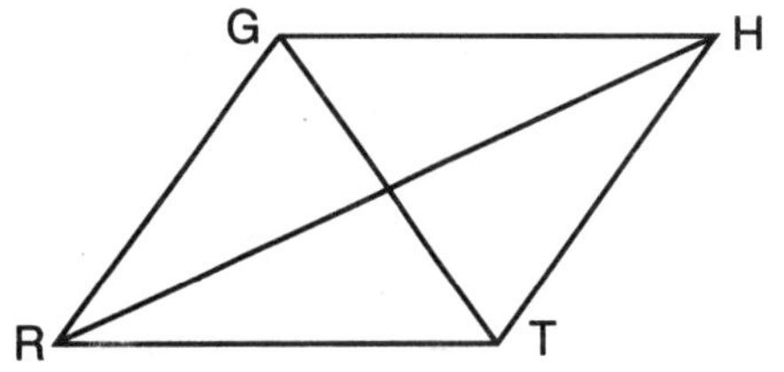

Ⓐ $\overline{HT}$

Ⓑ $\overline{GT}$

Ⓒ $\overline{RH}$

Ⓓ $\overline{RT}$

6 Pick one question from this test. Explain how and why you chose your answer.

____

____

____

____

____

____

Freshtown
新兴小镇

Directions: Look carefully at the picture below to locate various shapes.

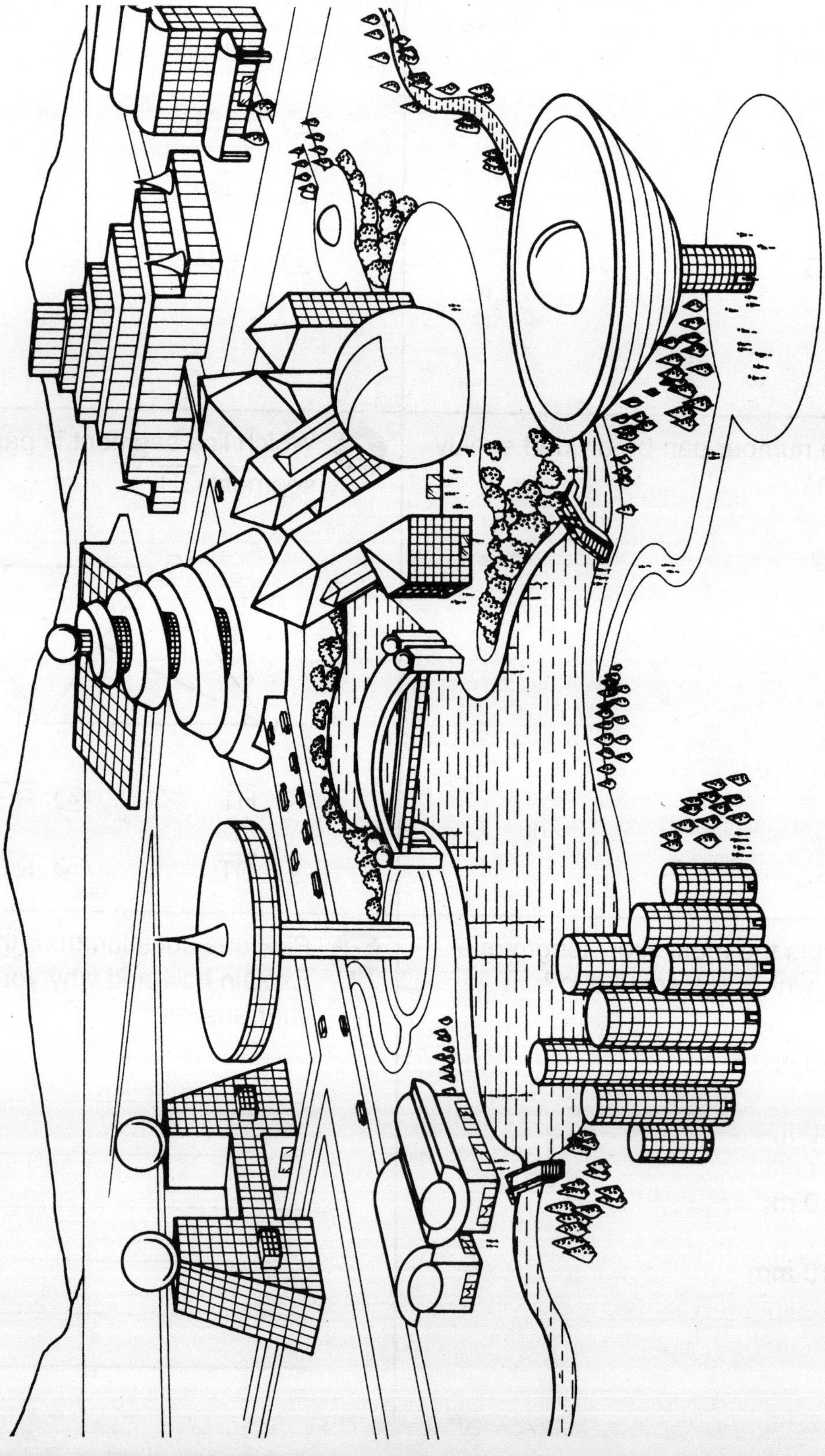

Properties Problems
描述特性 练习

Directions: Find four different-looking triangles in "Freshtown." Draw each triangle below. Then, on a separate sheet of paper, describe the properties of each triangle below.

Directions: Fill in the properties of shapes on this table.

3-D Shape (三维图形)	Number of Faces (面)	Number of Vertices (顶点)	Faces + Vertices	Number of Edges (边)
1. cube（立方体）				
2. rectangular prism（长方体）				
3. tetrahedron（四面体）				
4. square-based pyramid（方锥体）				
5. cone（圆锥体）				
6. triangular prism（三棱柱）				
7. hexagonal prism（六棱柱）				

Challenge

Directions: Look closely at the numbers in the *Faces + Vertices* and the *Number of Edges* columns. Do you notice a pattern?

1. Describe the relationship between these two columns. ____________________

2. The relationship between faces and the vertices and edges was discovered by Leonhard Euler (pronounced "oiler") and is called Euler's relation（欧拉关系）. Does it always work? ____________________

Acting It Out or Using Concrete Materials Group Problems

演示或使用小玩具解题 分组练习

Directions: Solve Problem A and show your work. Using the strategy you've studied this week, write and solve your own problems for Problem B and Problem C.

Problem A

Aimee has 10 containers in a row. She is putting blocks into the containers.

- Into every second container, she puts one block.
- Into every third container, she puts one block.
- Into every fourth container, she puts one block.
- Into every fifth container, she puts one block.
- Into every odd-numbered container, she puts two blocks.
- Into every even-numbered container, she puts three blocks.

1. How many blocks are in each container? ____________________
2. How many blocks are there in all? ____________________
3. Which containers have the most blocks? ____________________
4. Which containers have the fewest blocks? ____________________

Problem B: ____________________

____

____

____

____

____

Show work:

Answer: ____________________

Problem C: ____________________

____

____

____

____

____

Show work:

Answer: ____________________

Standardized Test Preparation 25
标准考前测试 25

1 Determine the next two numbers in this sequence.

2, 4, 9, 11, 16, ___, ___

Ⓐ 18, 21

Ⓑ 21, 23

Ⓒ 18, 24

Ⓓ 18, 23

2 Which 3-dimensional figure has only square faces?

Ⓕ rectangular prism

Ⓖ triangular prism

Ⓗ square-based pyramid

Ⓙ cube

3 Which number is both a mode and the median in this set of numbers?

(6, 12, 8, 9, 6, 15, 7, 8, 10, 3, 8)

Ⓐ 9

Ⓑ 6

Ⓒ 8

Ⓓ 12

4 What is the area of the rectangle?

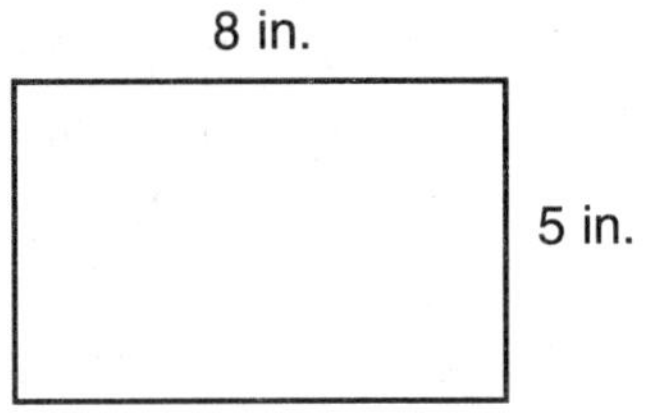

Ⓕ 40 in.^{2}

Ⓗ 168 in.^{2}

Ⓖ 52 in.^{2}

Ⓙ 26 in.^{2}

5 Which geometric shape will be created using reflective symmetry to complete it?

Ⓐ pentagon

Ⓑ hexagon

Ⓒ octagon

Ⓓ heptagon

6 Pick one question from this test. Explain how and why you chose your answer.

____

____

____

____

____

____

Investigate It

测一测

Directions: You need a coin and a centimeter (厘米) ruler to perform this experiment (实验). Place the coin at the edge of a desk or table so that it just hangs over the side. Use the flat (平面) of your palm (手掌) to shove the coin across the table, and then measure how far it travels. Repeat this nine more times, recording each measurement in the chart below. When finished, look at your results and answer the questions at the bottom.

Shoves (推)	Distance (cm) (距离)
1	
2	
3	
4	
5	
6	
7	
8	
9	
10	

1. What was the furthest distance traveled? ______________

2. What was the shortest distance? ______________

3. What was the range? ______________

4. Order your distances. What was the median distance? ______________

5. Was there a mode in your distances? If so, what was it? ______________

6. Plot your results on the following graph:

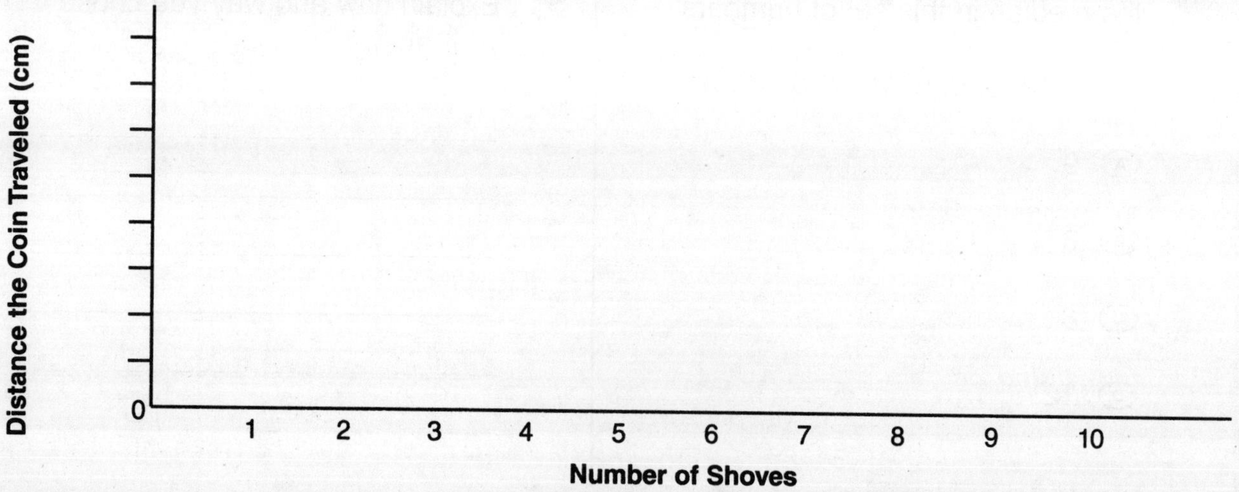

Using Simpler Numbers

化简数字 解题

What can you do if you come across（遇到） a problem that seems too difficult to solve? If it has large numbers or complicated（复杂的） number concepts（概念）, you can use simpler numbers to help you understand what you need to do. Then, you'll be ready to tackle（操作） the hard problem!

Using simpler numbers will help you understand what operations（处理） you need to use to solve the problem. Try replacing the large numbers in the problem with smaller numbers. Then, solve the problem. If the answer makes sense for the smaller numbers, then you can use the same operations with the larger numbers.

Another way to use simpler numbers is to break down the problem into smaller parts. As you solve each part, keep track of your answers by drawing pictures or a table. Soon, you may see a pattern that will help you solve the big problem.

Find a Starting Point（起始点）

Begin by using simpler numbers to work out a solution. You may find the operations and pattern that can be used to solve the more complicated problem.

Problem: Time for Art!

The Problem

A total of 16 artists worked for 10 hours to paint half of a mural（壁画）. Only 4 artists could stay to finish the other half. How long did it take the 4 artists to complete the other half?

Understanding the Problem

- *What do we know?*

  It took 16 artists 10 hours to paint half a mural. Four artists finished the other half of the mural.

- *What do we need to find out?*

  How long it took the 4 artists to complete the other half.

Planning and Communicating a Solution

- *Start with a Simpler Example*

  If it takes 4 artists 8 hours to paint half a mural, how long will it take 2 artists to paint the other half?

  First, find out how long it would take 1 artist to paint half the mural alone. He would have to work 4 times longer to do the job of the original four artists, so he would take 32 hours.

  $4 \times 8 = 32$ hours

  If 2 artists work on the other half, each will have to work half as much time as 1 artist. So, it would take them 16 hours.

  $32 \div 2 = 16$ hours

- *Solve the Original Problem*

  Start by working out how long it will take 1 artist to complete the work. We know that 16 artists take 10 hours, so 1 artist would have to work 16 times longer to do the job of 16 artists. So, 1 artist would take 160 hours.

  $16 \times 10 = 160$ hours

  If 4 artists work on the other half, they would each have to work only one-fourth as much time as 1 artist.

- *Do you see the answer?*

  They would each work 40 hours.

  $160 \div 4 = 40$ hours

Reflecting and Generalizing

By starting with a simpler example, you are able to solve the problem with more ease.

Standardized Test Preparation 26

标准考前测试 26

1 Dina's parents decided to make a garden in the backyard. If the space is 11 feet long and 4 feet wide, what is the area of the garden?

Ⓐ 23 ft.^{2}

Ⓑ 44 ft.^{2}

Ⓒ 69 ft.^{2}

Ⓓ 70 ft.^{2}

2 Which of the following weighs the greatest?

Ⓕ 0.54 kg

Ⓖ 0.499 kg

Ⓗ 0.6 kg

Ⓙ 0.132 kg

3 Which of these units of measurement would be used to measure the exact length of an insect leg?

Ⓐ feet

Ⓑ meter

Ⓒ millimeter

Ⓓ inch

4 What are the coordinates for point H?

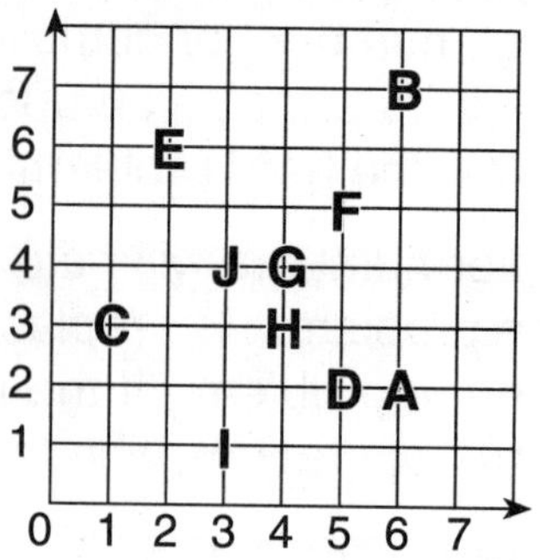

Ⓕ (3, 4)　　Ⓗ (4, 3)

Ⓖ (5, 3)　　Ⓙ (3, 3)

5 Find the mean in this set of numbers.

(4, 7, 9, 8, 4, 5, 5)

Ⓐ 3

Ⓑ 6

Ⓒ 4

Ⓓ 9

6 Pick one question from this test. Explain how and why you chose your answer.

____

____

____

____

____

____

Shape Cards

图形卡

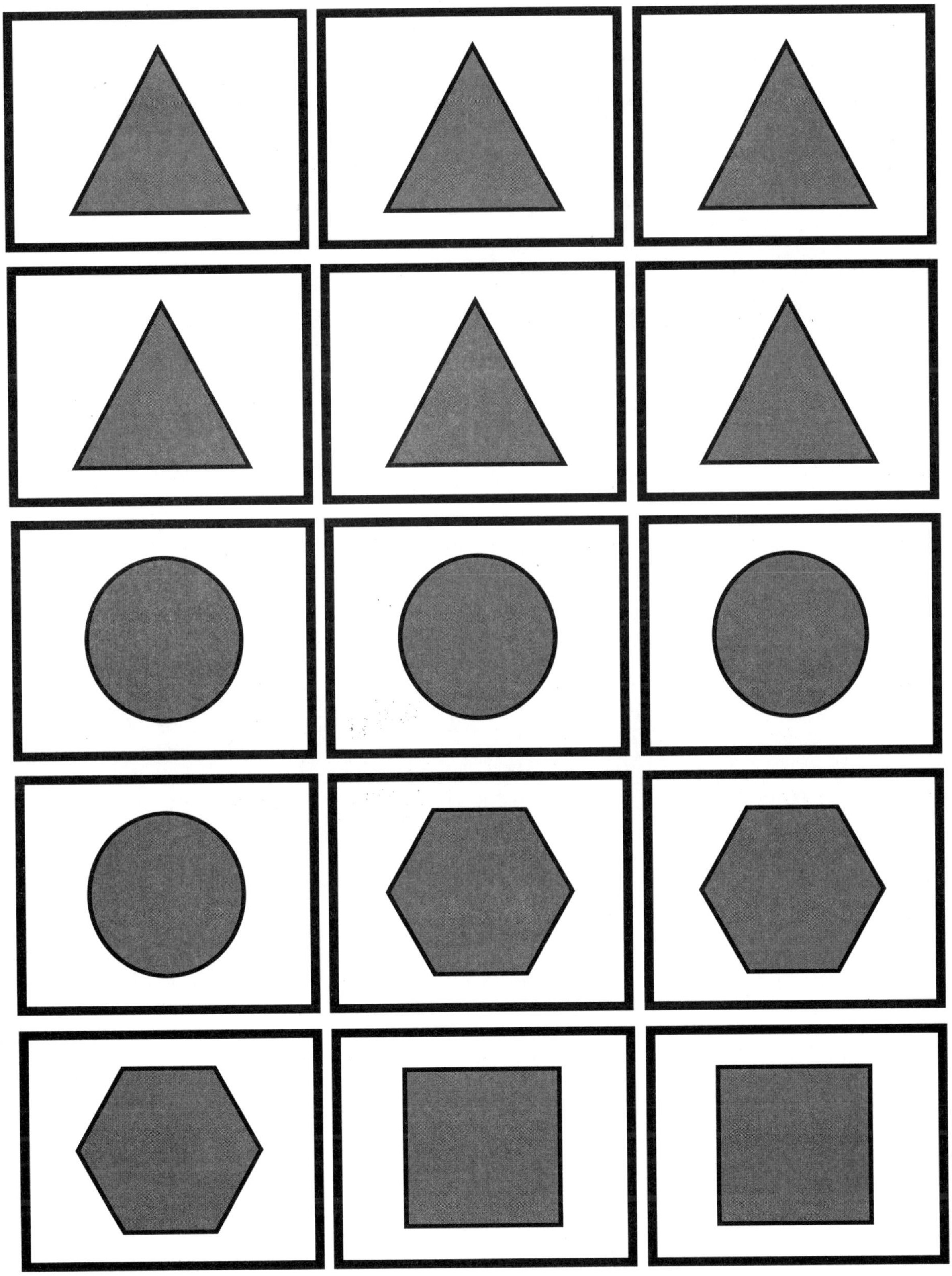

Shape Cards *(cont.)*

This page is left blank so you can cut out the manipulatives on the previous page.

Probability of Shapes

图形抽中的概率

Directions: Complete the graph below using the tally chart you created.

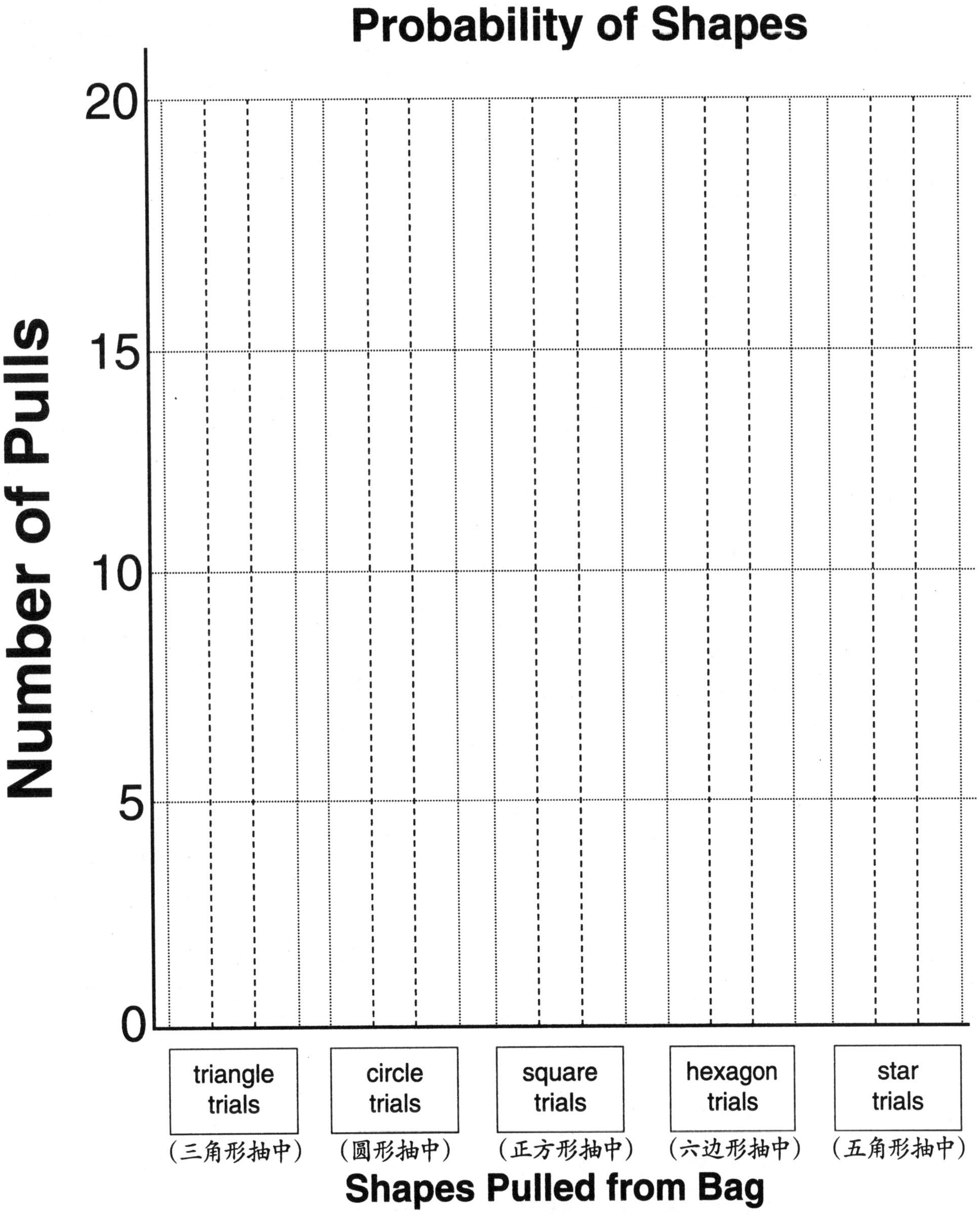

What Does an Architect Do?

建筑师是做什么的？

Who designs a new home? Who makes plans for a new skyscraper (摩天大楼)? An architect does. Architects decide what buildings will look like and how they will be built. They make drawings, or plans, that show how to build structures. They also decide what rooms the buildings will have.

Architects have the task of creating attractive, safe buildings for people. Many times, architects will design a few houses and a company will pay for these various designs to be built in one area. This is called a housing tract (房区). Often, in a housing tract, there will be three or four floor plans or designs. Then, the builders add something a little different to each house or the painters paint them differently so that all the houses don't look exactly the same.

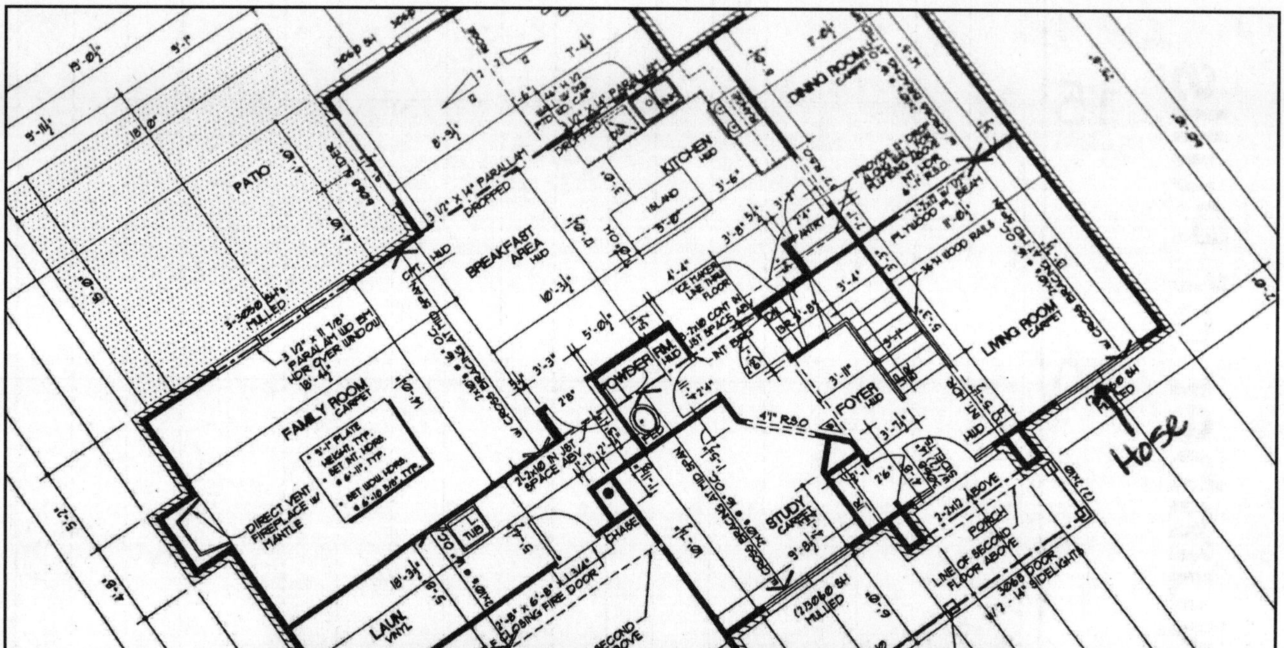

What Does an Architect Do? *(cont.)*

What Is the Problem?

Imagine that you are studying a housing tract for patterns. Use the information from the previous page and in the problems below to answer the questions. Decide if you have to use simpler numbers or break the problem down into smaller pieces.

Before you begin solving each problem, think about what information is given and what information you will need to figure out in order to complete the problem.

Remember: If you have trouble planning ways to solve each problem, reread it and rethink your strategy. When you have a solution, think about whether your answer makes sense.

Problem-Solving Strategy: Using Simpler Numbers

Problem A

If the houses in the housing tract are numbered from 1 to 150, how many houses will have the digit 8 as part of their house addresses?

Hint: Begin by breaking down the problem in smaller parts. How many 8s are in the house numbers from 1 to 9? How many 8s are in the house numbers from 10 to 19? Continue this process through 150? Put your results in a table with the following headings: House Numbers (1–9, 10–19, etc.), Eights in Ones Place, Eights in the Tens Place. Because it asks for houses—make sure that your chart doesn't count the double 8s as two houses in house #88.

Problem B

How many houses in a housing tract numbered 1–200 would have the digit 8 as part of the house address?

Hint: Continue adding to the above table to find your answer.

Problem C

How often will 3 appear in the sequence of numbers from 1 to 150? How often will 1 appear in the sequence of numbers from 1 to 150?

Extension

How does Problem A's answer change if you are counting 1s instead of 8s?

Hint: Add a hundreds column to your table. Make sure that you don't count double 1s, such as 11 or 101. Also, make sure you don't count the triple 1s in house #111.

Standardized Test Preparation 27

标准考前测试 27

1 Amy counted the number of petals on different flowers: 12, 23, 12, 9, 6, 13, 12, 14, 5, 6, 12. What is the median?

(A) 12

(B) 13

(C) 6

(D) 14

4 What is the volume of a rectangular prism with a length of 2.5 meters, a height of 9 meters, and a width of 10 meters?

(F) 21.5 meters cubed

(G) 225 meters squared

(H) 225 inches cubed

(J) 225 meters cubed

2 Sean has a square box for all his baseball cards. What is the area of the top of the lid if each side measures 12 inches?

(F) 47 $in.^2$

(G) 67 $in.^2$

(H) 144 $in.^2$

(J) 160 $in.^2$

5 Which of the answers is a multiple of these factors?

1, 2, 4, 7, 14

(A) 28

(B) 21

(C) 14

(D) 27

3 How many hours did Amy read?

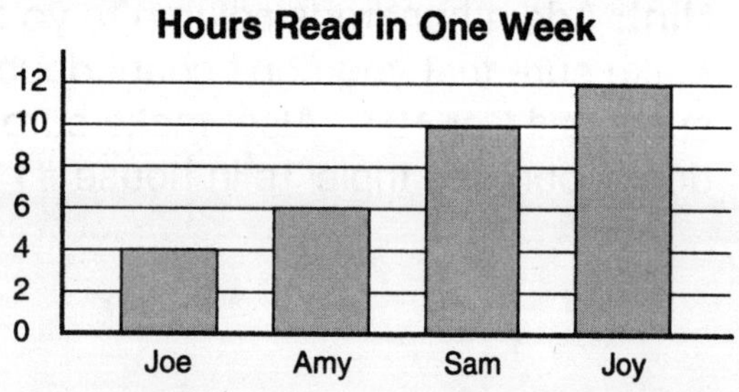

(A) 4 hours

(C) 5 hours

(B) 6 hours

(D) 7 hours

6 Pick one question from this test. Explain how and why you chose your answer.

____

____

____

____

____

____

Marble Graph
弹珠图表

Directions: Use the coordinate points to complete the graph below.

Data Coordinates

(1, 4)	(6, 27)
(2, 9)	(7, 31)
(3, 13)	(8, 33)
(4, 18)	(9, 30)
(5, 21)	(10, 25)

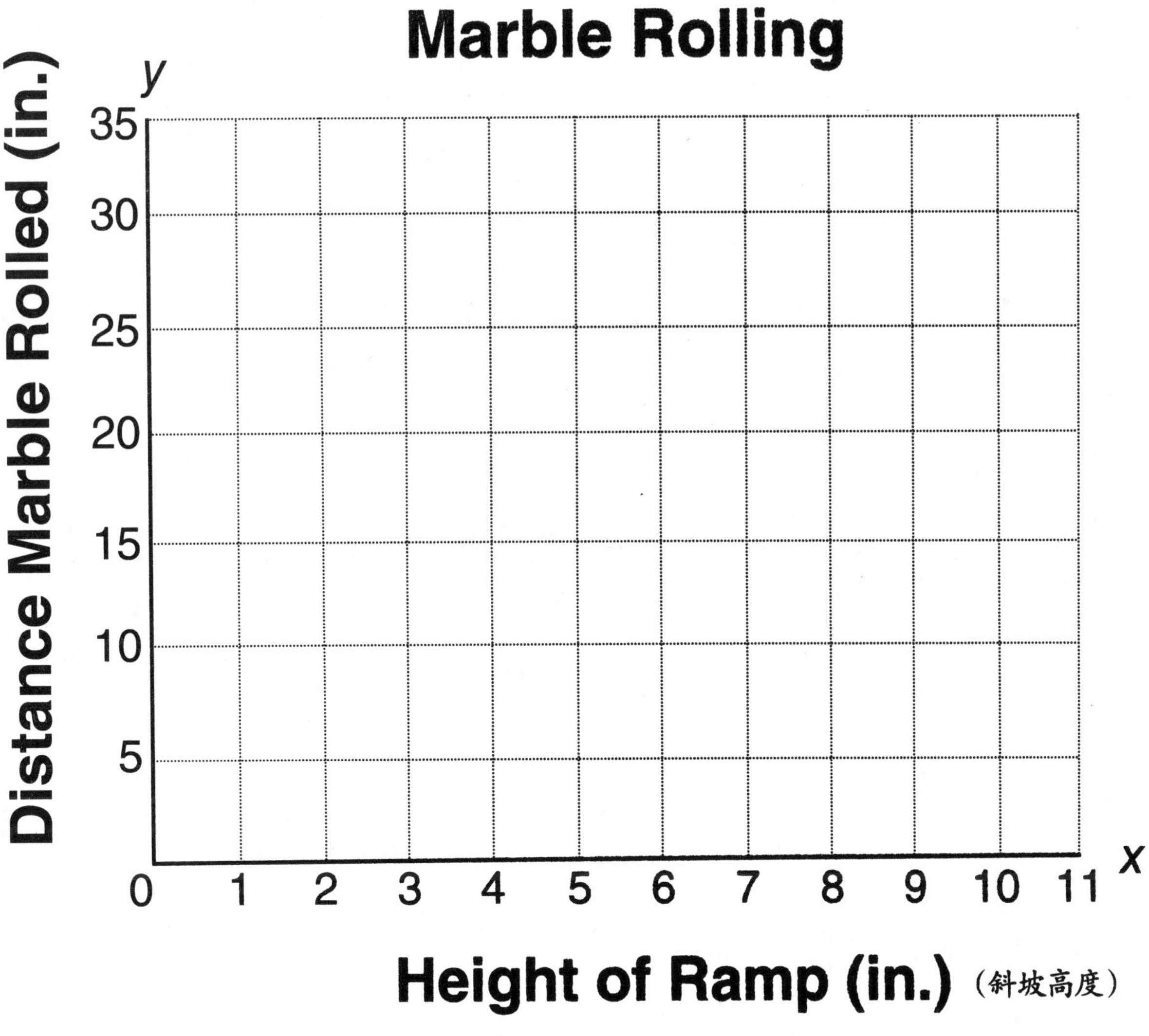

Marble Graph *(cont.)*

Section A

Directions: Answer the questions below by looking at the table of coordinates for the ramp experiment and the completed graph.

1. What can you tell by looking at this graph?

____

____

____

____

2. Why do you think the line on the graph starts to go back down after the ramp is 8 inches high?

____

____

3. If the *y*-axis is 11, what would be a good estimate for the *x*-axis? ____________

4. If the height of the ramp was $4\frac{1}{2}$ inches, how far do you think the marble would roll?

____

Section B

Directions: Write the letter that represents the point for each set of coordinates.

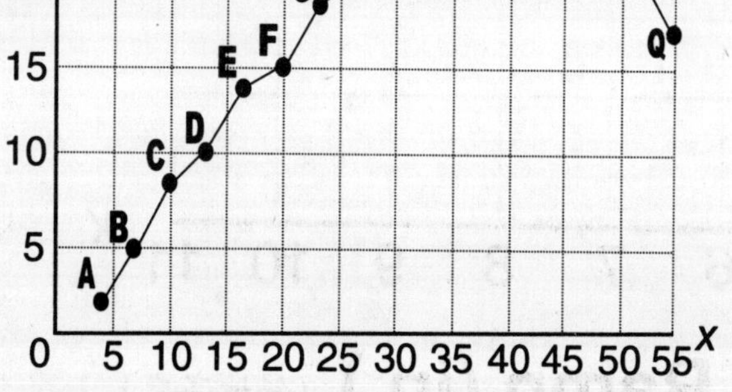

1. (7, 5) ____
2. (27, 20) ____
3. (20, 15) ____
4. (42, 29) ____
5. (30, 23) ____
6. (35, 28) ____
7. (4, 2) ____
8. (45, 27) ____
9. (48, 25) ____
10. (55, 17) ____
11. (10, 8) ____
12. (17, 13) ____
13. (38, 30) ____
14. (50, 22) ____
15. (13, 10) ____
16. (23, 18) ____

Using Simpler Numbers Problems

化简数字解题 练习

Directions: Solve each problem and show your work.

Problem A

How many 1" cubes does it take to fill this small box, which is about the size of a pencil box or a shoe box?

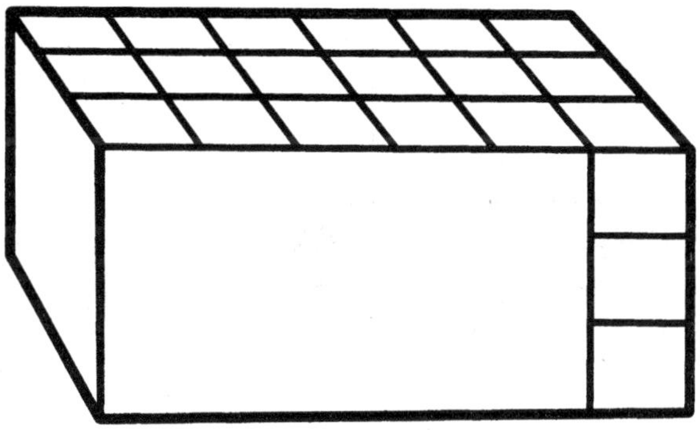

Problem B

1. Which is more—4^5 or 4 x 5?
2. Which is more—3^4 or 3 x 4?
3. Which is more—2^6 or 2 x 6?
4. Which is more—5^2 or 5 x 2?
5. Which is more—6^3 or 6 x 3?

Standardized Test Preparation 28
标准考前测试 28

1 Which of these letters illustrates reflective symmetry?

Ⓐ K

Ⓑ L

Ⓒ A

Ⓓ S

2 The roots of a redwood tree（红杉树）go 100 feet deep. Which is the closest measurement?

Ⓕ 30 yd. 4 ft.

Ⓖ 102 ft. 9 in.

Ⓗ 35 yd. 1 ft.

Ⓙ 33 yd. 1 ft.

3 Solve.

$379 \div 7 = \_\_\_$

Ⓐ 54 R1

Ⓑ 55 R1

Ⓒ 59 R1

Ⓓ 54 R2

4 Which polyhedron（多面体）has five faces, four of them triangles?

Ⓕ octahedron（八面体）

Ⓖ tetrahedron（四面体）

Ⓗ triangular prism（三棱柱）

Ⓙ square-based pyramid（方锥体）

5 How do you find the area of a rectangle?

Ⓐ perimeter x length

Ⓑ length + width

Ⓒ length x width

Ⓓ perimeter x height

6 Pick one question from this test. Explain how and why you chose your answer.

____

____

____

____

____

____

Endangered Species
濒临灭绝的物种

Graph A

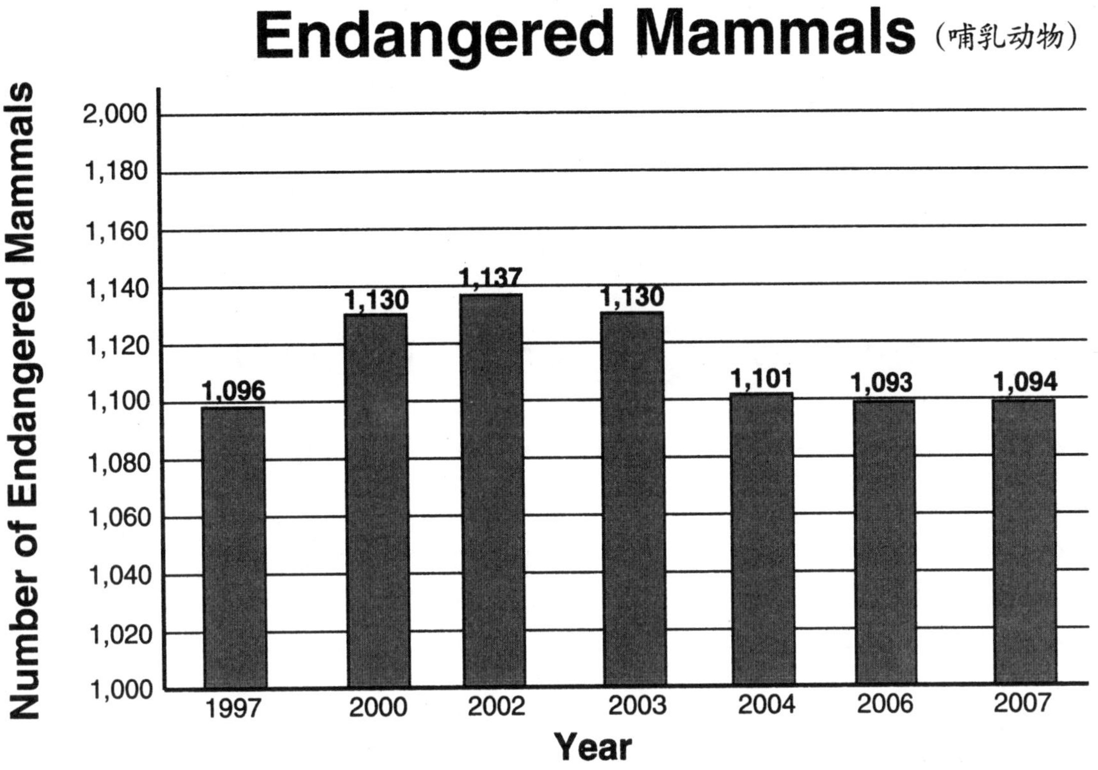

Graph B

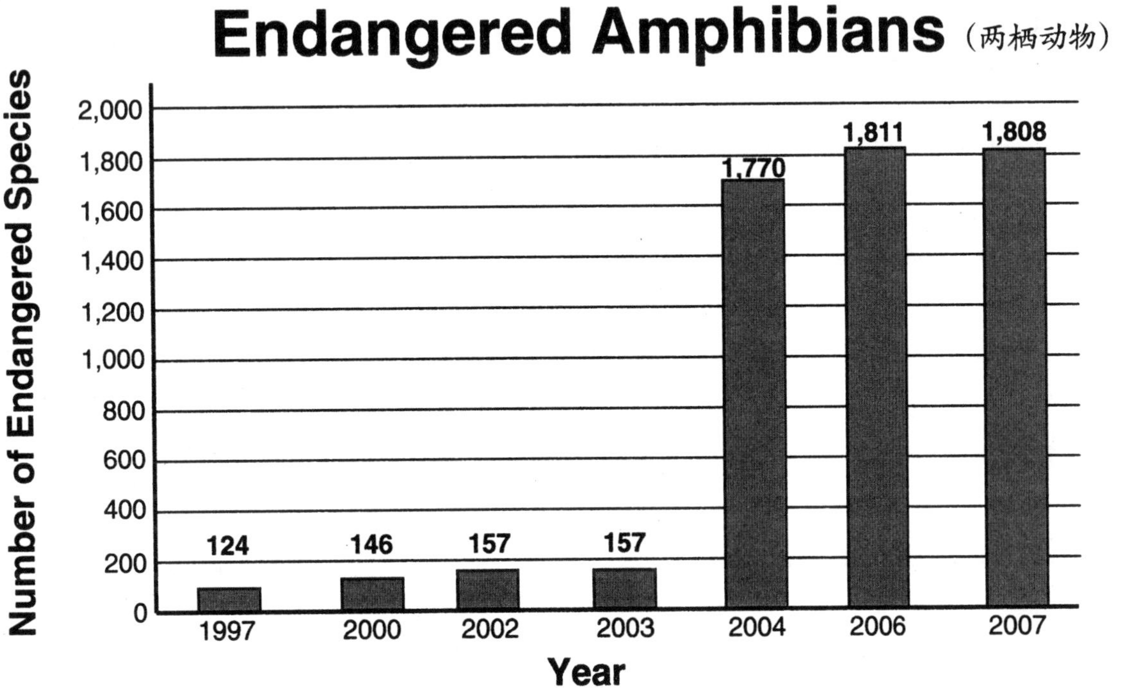

Endangered Reptiles

濒临灭绝的爬行动物

Directions: Complete the graph using the data below.

Endangered Reptiles Table

Year	1997	2000	2002	2003	2004	2006	2007
Endangered Species	253	296	293	293	304	341	422

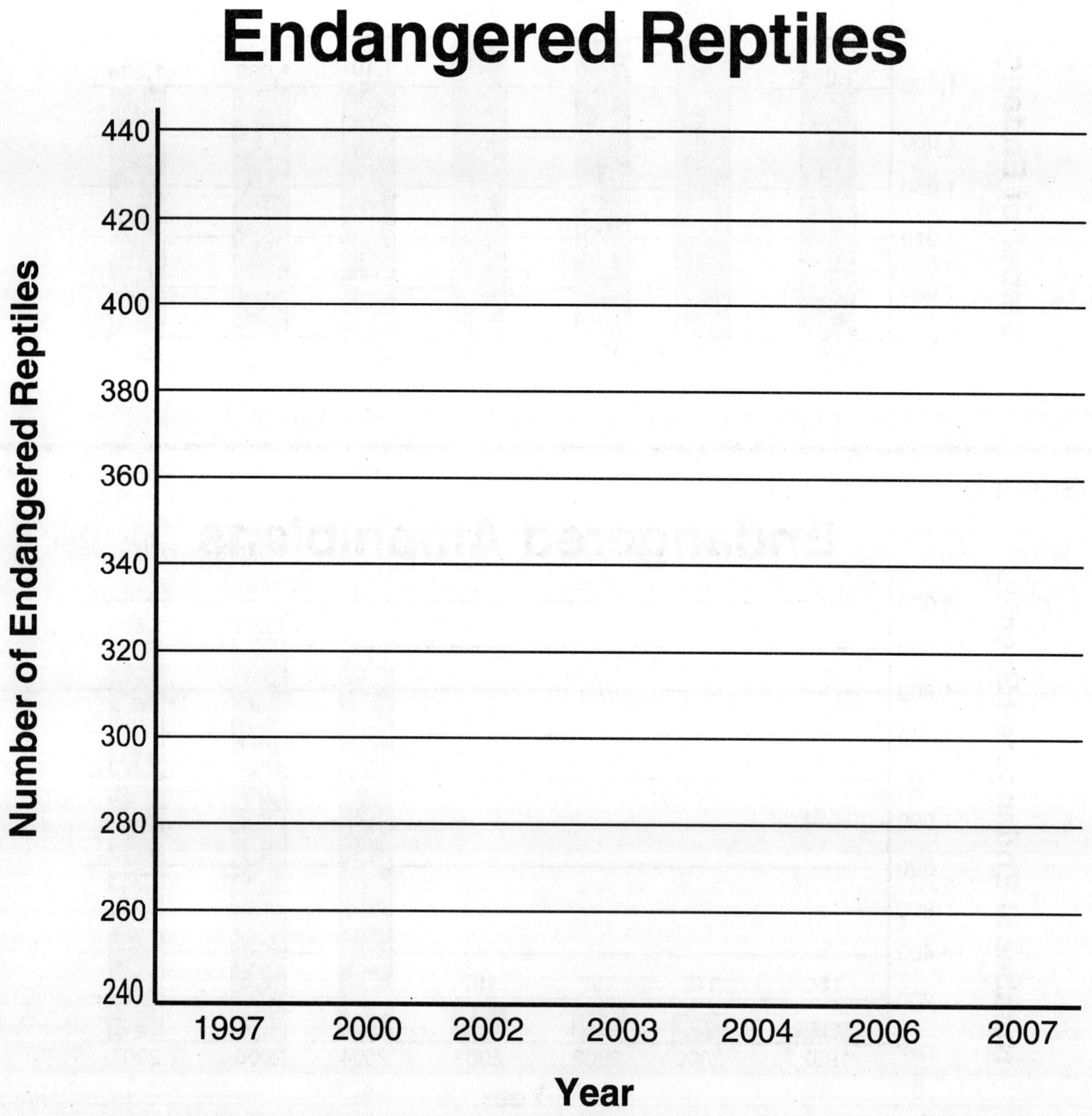

Endangered Reptiles *(cont.)*

Directions: Answer the following problems.

1. What can you tell by reading the Endangered Reptiles graph?

____

____

____

____

2. How many more reptiles were on the endangered list in 2007 than 2000? ______________

3. How many more amphibians were endangered in 2007 than reptiles in the same year?

____

4. How many species of amphibians and reptiles were endangered in 2006? ______________

5. How many species of mammals and amphibians were endangered in 2003? ____________

6. How many total species of mammals, reptiles, and amphibians were endangered in 2007?

____

7. In the remaining space, write questions of your own that can be answered using the information in the graphs.

____

____

____

____

____

____

____

____

____

____

The Floors Under Your Feet
你脚下的地板

Look down at the floor. What do you see? Some floors are covered in carpet（地毯）. Other floors are made of wood. Some floors have tiles（瓷砖） on them. Many tiles are squares. They may be big or small.

Some tile floors have a checkerboard（棋盘格） pattern with different colors. Other floors have tiles of the same color. Some floors have tiles arranged in patterns. Tiles can make a floor look more interesting.

Sometimes the tiles in the floor make a pattern, but other times they make a picture. Tile floors can show pictures of flowers or even people! When the builder makes a tile floor with a pattern or picture, the builder has to be very careful. The builder has to make sure there are enough tiles of each color to finish the floor. So the next time you walk down a hallway or into a building, be sure to check out the floor!

The Floors Under Your Feet *(cont.)*

What Is the Problem?

Imagine that you have to order the tiles you want to cover your floor. You will have to know how many tiles you will need before you order them.

Use the information above and in the problems below to answer the questions. Decide if you have to use simpler numbers or break the problem down into smaller pieces.

Before you begin solving each problem, think about what information is given and what information you will need to figure out in order to complete the problem.

Remember: If you have trouble planning ways to solve each problem, reread it and rethink your strategy. When you have a solution, think about whether your answer makes sense.

Problem-Solving Strategy: Using Simpler Numbers

Problem A

A dining room（餐厅）has dimensions（规模）of 15 feet by 12 feet. A tile is 2 square feet. How many tiles do you need for the floor?

Problem B

A kitchen（厨房）is 25 feet long and 18 feet wide. A tile is 2 square feet. How many tiles do you need for the floor?

Problem C

It took you 9 hours to tile the dining room floor. At this rate, how long will it take you to tile the kitchen floor?

Extension

Measure your classroom floor. What is the perimeter? What is the area? How many tiles will you need to cover the floor if they are each 2 feet long by 2 feet wide? Use simpler numbers to help you find the answer. Draw a picture of your new floor.

Standardized Test Preparation 29

标准考前测试 29

1 Which of these answers expresses a fraction in simplest form?

Ⓐ $\frac{2}{8}$

Ⓑ $\frac{9}{21}$

Ⓒ $\frac{8}{11}$

Ⓓ $\frac{12}{32}$

2 What is the area of the rectangle?

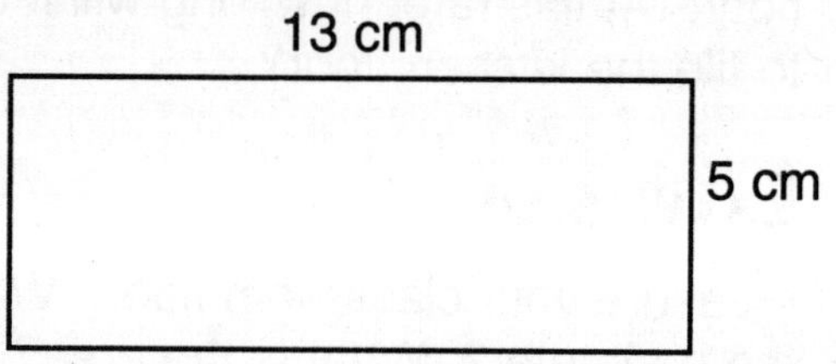

Ⓕ 36 cm^2　　Ⓗ 55 cm^2

Ⓖ 45 cm^2　　Ⓙ 65 cm^2

3 What are the coordinates for point I?

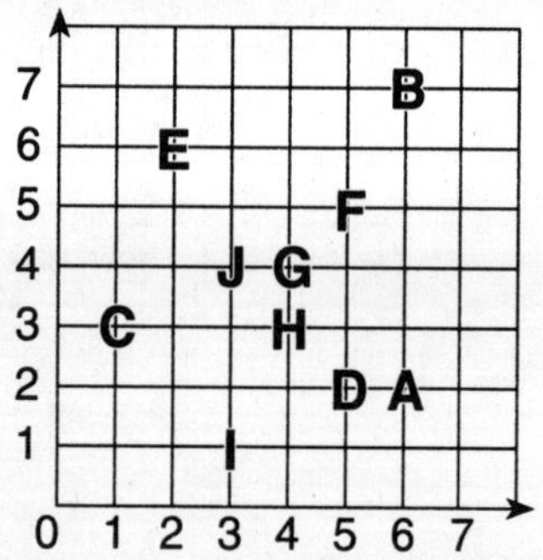

Ⓐ (2, 4)　　Ⓒ (3, 1)

Ⓑ (1, 3)　　Ⓓ (4, 7)

4 Triangles ABC and DEF are congruent. Which of the following is true?

Ⓕ The sides are equal, but the angles are unequal.

Ⓖ The sides are unequal.

Ⓗ The angles and sides are similar.

Ⓙ The angles and sides are equal.

5 What was the average monthly temperature for May as shown on the line graph?

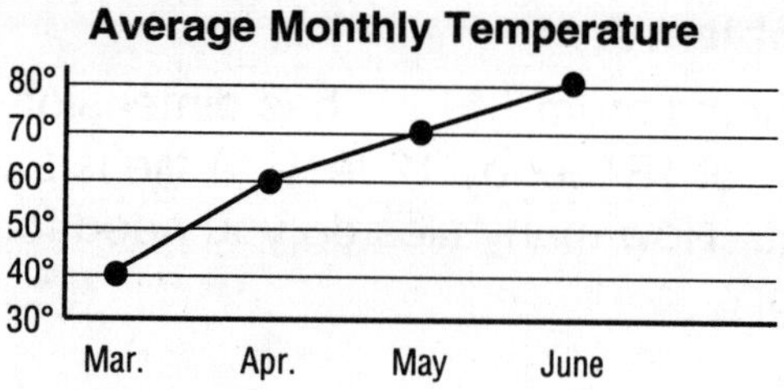

Ⓐ 50°　　Ⓒ 60°

Ⓑ 70°　　Ⓓ 65°

6 Pick one question from this test. Explain how and why you chose your answer.

____

____

____

____

____

____

Lesson 30

Flipping Coins

抛硬币

Directions: Experimental probability tells you what actually happens when you flip a coin or roll a die（骰子）. What are the chances of flipping one coin and landing on heads? There are two possibilities, heads or tails (1 chance in 2).

1. Flip a coin 48 times and record the results below.

_____ _____ _____ _____ _____ _____ _____ _____

_____ _____ _____ _____ _____ _____ _____ _____

_____ _____ _____ _____ _____ _____ _____ _____

_____ _____ _____ _____ _____ _____ _____ _____

_____ _____ _____ _____ _____ _____ _____ _____

_____ _____ _____ _____ _____ _____ _____ _____

2. How many times did you flip heads? __________

These are the 4 possibilities when flipping two coins: (C = Coin, H = Heads, T = Tails)

C1	C2
H	H

C1	C2
H	T

C1	C2
T	H

C1	C2
T	T

3. What are the chances of flipping two heads? ________________

4. What are the chances of flipping two tails? ________________

5. What are the chances of flipping 1 head and 1 tail? ________________

Flipping Coins *(cont.)*

Directions: Experimental probability tells you what actually happens when you flip a coin or roll a die. What are the chances of flipping one coin and landing on heads（正面朝上落下）? There are two possibilities, heads or tails（反面）(1 chance in 2).

6. Flip two coins 48 times and record the results below.

______ ______ ______ ______ ______ ______ ______ ______

______ ______ ______ ______ ______ ______ ______ ______

______ ______ ______ ______ ______ ______ ______ ______

______ ______ ______ ______ ______ ______ ______ ______

______ ______ ______ ______ ______ ______ ______ ______

______ ______ ______ ______ ______ ______ ______ ______

7. How many times did you flip two heads? ____________________

8. How many times did you flip two tails? ____________________

9. How many times did you flip one head and one tail? ____________________

Will It or Won't It?

会还是不会？

Directions: Draw a big 0–1 probability scale（概率尺度）in the box.

You will need to use the fraction method（分数法）to record the probability for all of the following statements. Write the fraction in the spaces provided, placing the numbers for each of these events on your probability scale.

1. What is the probability that you toss a coin（抛硬币）and get tails? ________________

2. You have a bag of fruit with six apples, two bananas, and four plums（梅子）. What is the probability of your randomly（任意地）taking out
   - **a.** an apple? __________
   - **b.** a plum? __________
   - **c.** a banana? __________

3. In a group of 10 students, eight are wearing tennis shoes, one wears glasses, five have packed lunch（盒装午餐）, six are girls, and two like olives（橄榄）. Work out the probability of the following:
   - **a.** having packed a lunch ____________
   - **b.** being a boy ____________
   - **c.** not wearing glasses ____________
   - **d.** liking olives ____________
   - **e.** wearing tennis shoes ____________

4. If the probability of being late for school is $\frac{1}{2}$, what is the probability of being on time? ________________

5. You throw a die.
   - **a.** What is the probability of getting a 6? ___________
   - **b.** What is the probability of getting a number less than 4? ___________
   - **c.** What is the probability of getting an odd number? ___________

Will It or Won't It? *(cont.)*

Directions: Sometimes probability is estimated because it is impossible to be sure of the outcome. You have to think about how likely it is. Can you estimate a probability for these events? Can you place them on the probability scale?

6. Think of a main road near your school. What is the probability that within five minutes

a. a bus will come by? __________

b. most cars going by will be red? __________

Directions: Draw a big 0–1 probability scale in the box.

You will need to use the fraction method to record the probability for all of the following statements. Write the fraction in the spaces provided, placing the numbers for each of these events on your probability scale.

Directions: Here is the spelling of EXCELLENCE. Imagine you have the letters for the whole word on individual cards. You shuffle（洗牌） them and turn them over on a table.

7. What is the probability of turning over:

a. L? _______ **b.** C? _______ **c.** E? _______ **d.** X? _______ **e.** N? _______

Directions: Here is the spelling of MISSISSIPPI. Imagine you have the letters for the whole word on individual cards. You shuffle them and turn them over on a table.

8. What is the probability of turning over（翻转）:

a. M? _______ **b.** S? _______ **c.** I? _______ **d.** P? _______

Using Simpler Numbers Group Problems

化简数字解题 分组练习

Directions: Solve Problem A and show your work. Using the strategy you've studied this week, write and solve your own problems for Problem B and Problem C.

Problem A

Suzy is hopping and jumping（一蹦三跳） along the beach. She does three hops followed by one jump. From a standing start（原地出发）, how many footprints（脚印） in the sand would there be for her after 3 jumps? 10 jumps? 20 jumps?

(**Hint:** One foot is used when hopping and both feet are used when jumping.)

Problem B: ____________________

____

____

____

____

____

Show work:

Answer: ____________________

Problem C: ____________________

____

____

____

____

____

Show work:

Answer: ____________________

Standardized Test Preparation 30
标准考前测试 30

1 It took Elaine 7 hours and 44 minutes to drive from Los Angeles（洛杉矶）to San Jose（圣何塞）. She left at 9:15 a.m. When did she arrive in San Jose?

Ⓐ 4:49 p.m.

Ⓑ 4:59 a.m.

Ⓒ 5:50 p.m.

Ⓓ 4:59 p.m.

2 What are the coordinates for point J?

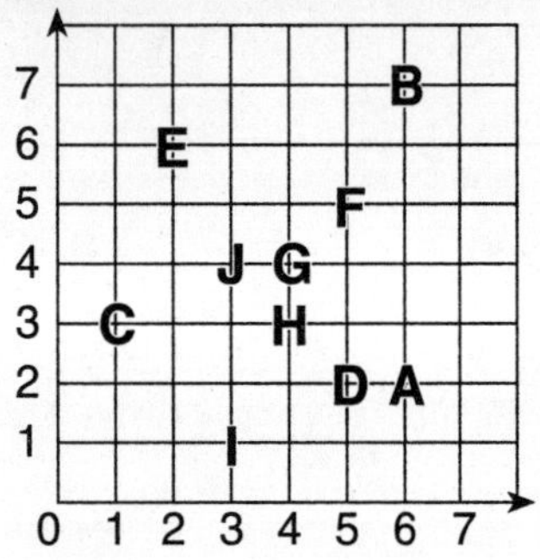

Ⓕ (4, 3)　Ⓗ (4, 4)

Ⓖ (3, 4)　Ⓙ (3, 3)

3 Determine the next two numbers in this sequence.

0, 1, 1, 2, 3, 5, 8, ___, ___

Ⓐ 13, 16

Ⓑ 12, 17

Ⓒ 13, 21

Ⓓ 13, 23

4 Which three-dimensional figure has eight triangular faces?

Ⓕ tetrahedron

Ⓖ octahedron

Ⓗ square-based pyramid

Ⓙ cube

5 The shape of a backyard is shown here. What is the perimeter of the backyard?

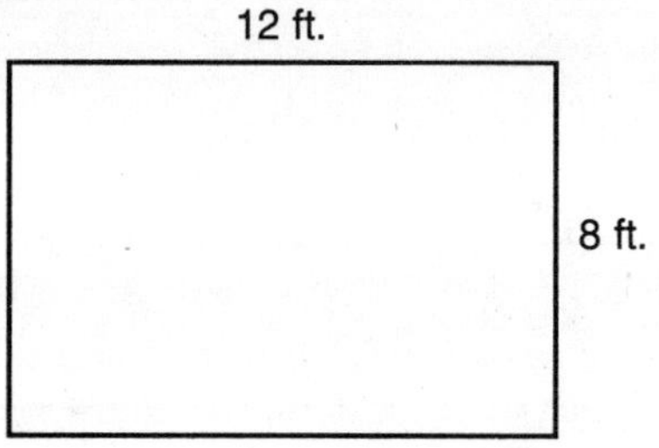

Ⓐ 15 ft.　Ⓒ 40 ft.

Ⓑ 23 ft.　Ⓓ 96 ft.

6 Pick one question from this test. Explain how and why you chose your answer.

____

____

____

____

____

____

Appendix A: 附录A

Game Directions 游戏指南

Match It! Directions

对对碰！游戏指南

Match It! Directions *(cont.)*

What You Need

- *It's . . . Totally Viewed* game board
- spinner（旋钮）(Divide the spinner into 4 parts and write the numbers 1–4 on it.)
- *Match It!* cards
- *Counters*（计数器） or something else to use as game markers
- pencils and paper

Object of the Game

- Match math problems and their answers. Use your memory to do this. Be the first player to move around the board and collect 50 points.

Setting Up the Game

- Place the game board in the middle of all the players.
- Shuffle the 36 cards and place them facedown. Arrange them in 6 rows with 6 cards in each row.
- Each player places a game marker on the space marked **START**.

How to Play the Game

- The youngest player goes first. Then, play passes to the left.
- For each turn, flip over two cards. Make sure that everyone can see the cards when you flip them.
- If you flip a mathematics problem, solve it. Then, try to find the correct answer. If you flip an answer, remember where it is and try to find the correct problem. You are trying to match problems with their answers.
- If you don't have a match, your turn is over.
- If you make a pair, spin the spinner and move that many spaces. You can move onto any space touching the space you are currently on. When you stop, record the number. You need to add your landing points together until you reach 50.
- Your turn is over after you move. Keep any pairs you find.

How to Win the Game

- The first player to land on spaces that add up to at least 50 wins!
- Second place goes to whoever has the most matches. (If the winner had the most matches, there is no second place!)

Find It! Directions

找找看！游戏指南

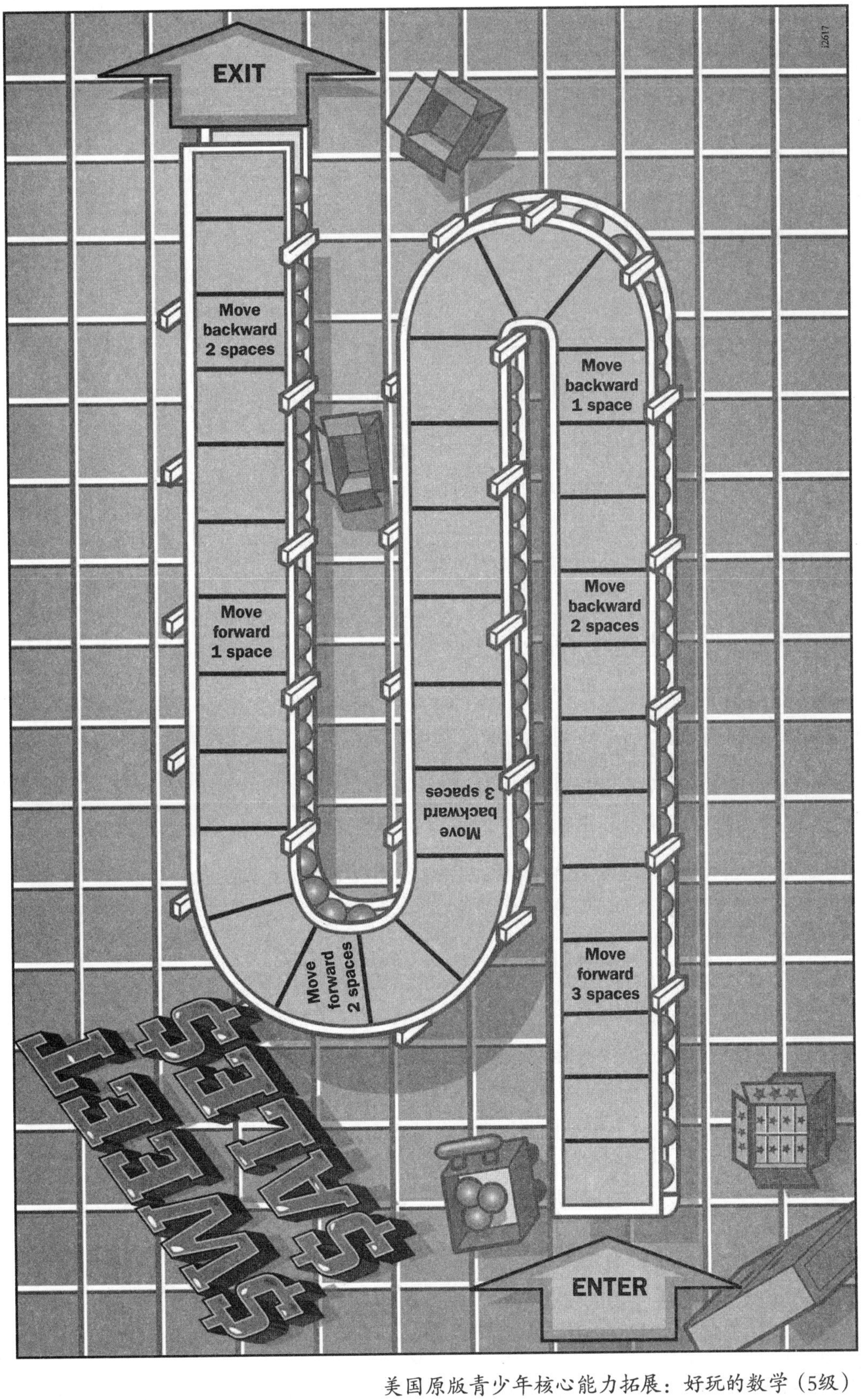

Find It! Directions *(cont.)*

What You Need

- *Sweet Sales* game board
- spinner (Divide the spinner into 4 parts and write the numbers 1–4 on it.)
- *Find It!* cards
- *Counters* or something else to use as game markers
- pencils and paper

Object of the Game

- Be the first player to move through the candy factory（糖果工厂）to the **EXIT**.

Setting Up the Game

- Place the game board in the middle of all the players.
- Shuffle the 18 question cards and place them facedown in a pile within easy reach of all the players.
- Shuffle the 18 answer cards and deal an equal amount to each player, using as many cards as possible.
- Each player places a game marker on **ENTER**.

How to Play the Game

- The player whose birthday is closest to today goes first. Then, play passes to the left.
- For each turn, flip over a question card.
- Everyone has to solve the problem.
- Whoever has the correct answer card spins the spinner and moves forward that many spaces. Follow any directions on the game board (move backward or move forward). Your turn is over after you move.
- Discard the question and answer cards in two separate piles. If you run out of cards before someone wins, shuffle the piles separately and use the cards again.
- Continue playing until someone gets through the candy factory!

How to Win the Game

- The first player to land on **EXIT** (an exact roll is not required) wins!

Solve It! Directions

答题王！游戏指南

Solve It! Directions *(cont.)*

What You Need

- *Highway 10* game board
- spinner (Divide the spinner into 4 parts and write the numbers 1–4 on it.)
- *Solve It!* Cards
- *Counters* or something else to use as game markers
- pencils and paper

Object of the Game

- Be the first player to travel 500 miles along Highway 10.

Setting Up the Game

- Place the game board in the middle of all the players.
- Shuffle the 18 cards and place them facedown in a pile within easy reach of all the players.
- Each player places a game marker on **START**.
- Decide whether to play the regular level or challenge level.

How to Play the Game

- The shortest player goes first. Then, play passes to the left.
- For each turn, draw the top card from the deck.
- Everyone solves the problem.
- Compare your answers. Decide who solved the problem correctly.
- Everyone who has solved the problem correctly gets to spin the spinner and move. You can move in any direction on the board. Once you start in one direction with a spin, you must continue that way.
- You cannot earn the same mileage（英里数）more than once. So, if you land on "150 miles" once, that's it. You cannot earn 150 miles again.
- **Regular level:** You can only move your piece.
- **Challenge level:** You can choose to not move your piece and instead move another player's piece the number you spun. This is a way to make other players lose miles.
- Place the card in a discard pile. Once you run out of cards, shuffle the discard pile and use the cards again.

How to Win the Game

- The first player to travel a total of 500 miles along Highway 10 wins!

Rewrite It! Directions
变一变！游戏指南

Rewrite It! Directions *(cont.)*

What You Need

- *Save the Princess* game board
- spinner (Divide the spinner into 2 parts and write the numbers 1–2 on it.)
- *Rewrite It!* Cards
- *Counters* or something else to use as game markers
- pencils and paper

Object of the Game

- **Regular level:** Be the first player to climb up the ladder（梯子） and reach the princess（公主）.
- **Challenge level:** Be the first player to climb up the ladder and save the princess by bringing her back down the ladder.

Setting Up the Game

- Place the game board in the middle of all the players.
- Shuffle the 18 cards and place them facedown in a pile within easy reach of all the players.
- Each player chooses a colored ladder and places a game marker on the word **Start** in the grass.
- Decide whether to play the regular level or challenge level.

How to Play the Game

- The oldest player goes first. Then, play passes to the left.
- For each turn, draw the top card from the deck.
- Everyone has to rewrite the problem in some other way. You can do this in any way you want. Some possibilities are to use inverse operations, add variables, create word problems, or apply mathematical properties.
- Have the group decide which rewritten problems are accurate (i.e., same problem rewritten into a new form; you haven't changed the problem).
- Everyone who has rewritten the problem correctly gets to spin the spinner and move.
- **Challenge level:** You have to rewrite the problem in a different way every turn. You cannot repeat ways of writing the problems (i.e., the first problem is rewritten using inverse operations; the second problem is rewritten using a word problem; the third problem is rewritten using variables).
- Place the card in a discard pile. Once you run out of cards, shuffle the discard pile and use the cards again.

How to Win the Game

- **Regular level:** The first player to land the princess wins!
- **Challenge level:** The first player to land on the princess and climb back down the ladder by rewriting problems wins!

Appendix B: 附录B

Mathematics Chart

数学用表

Three-Dimensional Shapes（三维图形）

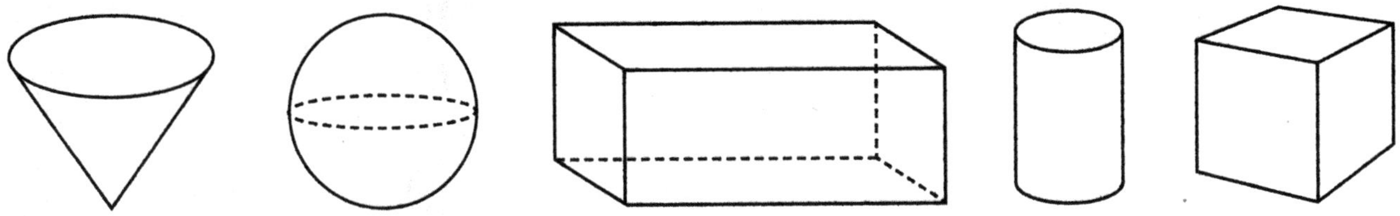

cone（圆锥体） sphere（球体） rectangular prism（方棱柱） cylinder（圆柱体） cube（立方体）

Two-Dimensional Shapes

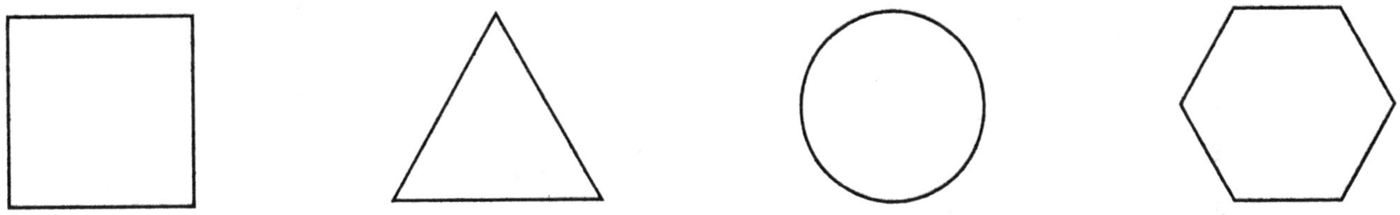

square（正方形） triangle（三角形） circle（圆形） hexagon（六边形）

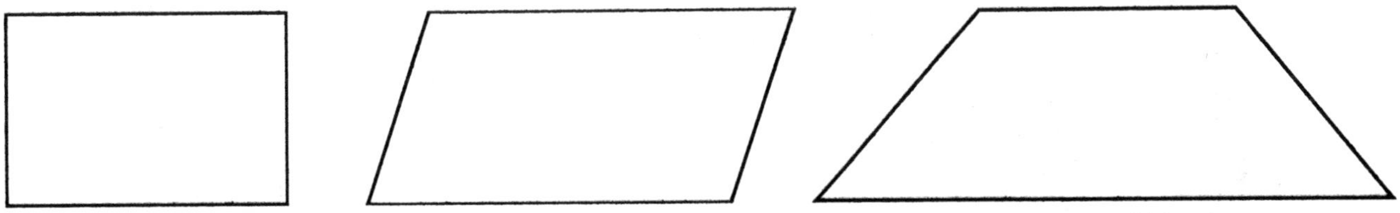

rectangle（矩形） rhombus（平行四边形） trapezoid（梯形）

Measurement（测量）

Perimeter（周长）

square $P = 4s$

rectangle $P = 2l + 2w$ or $P = 2(l + w)$

Area（面积）

square $A = s^2$

rectangle $A = lw$ or $A = bh$

Volume（体积）

cube $V = s^3$

rectangular prism $V = l \times w \times h$

Mathematics Chart *(cont.)*

Time（时间）

1 year（年）= 365 days
1 year = 12 months（月）
1 year = 52 weeks（周）
1 week = 7 days（日）
1 day = 24 hours（小时）
1 hour = 60 minutes（分钟）
1 minute = 60 seconds（秒）

Money（货币）

1 dollar（美元） = 100 pennies（美分）
1 dollar = 4 quarters（25美分）
1 dollar = 10 dimes（10美分）
1 dollar = 20 nickels（5美分）
1 quarter = 25¢
1 dime = 10¢
1 nickel = 5¢
1 penny = 1¢

Weight（重量）

Metric（公制）

1 kilogram（千克）= 1,000 grams（克 ）
1 gram = 1,000 milligrams（毫克）

Customary（惯例法）

1 ton（吨）= 2,000 pounds（磅）= 1000 kg
1 pound = 16 ounces（盎司）= 0.4536 kg

Length（长度）

Metric

1 kilometer（千米）= 1,000 meters（米）
1 meter = 100 centimeters（厘米）
1 centimeter = 10 millimeters（毫米）

Customary

1 mile（英里）=5,280 feet（英尺）=1609.344 meters
1 yard（码）= 3 feet = 0.9144 meters
1 foot = 12 inches（英寸）= 0.3048 meters

Fractions（分数）

$\frac{1}{2}$ ← numerator（分子）, ← denominator（分母）

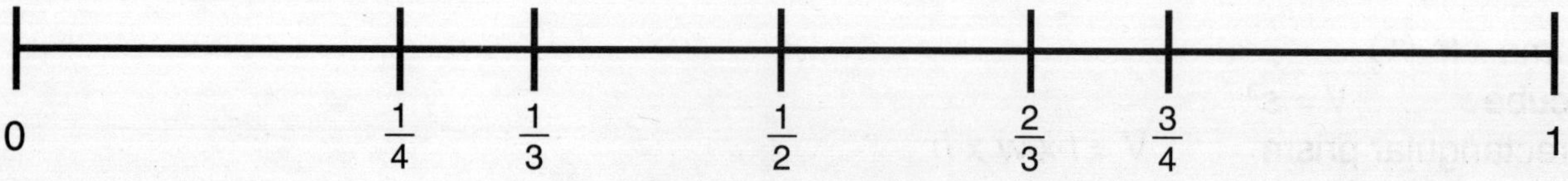

Mathematics Chart *(cont.)*

Gallon Conversions（加仑换算表）

Key

G = gallon（加仑）

Q = quart（夸脱）

P = pint（品脱）

C = cup（量杯）

Capacity and Volume（容积和体积）

Metric

1 liter（升）= 1,000 milliliters（毫升）

Customary

1 gallon = 4 quarts = 3784 milliliters

1 quart = 2 pints = 1101 milliliters

1 pint = 2 cups = 568.26125 milliliters

1 cup = 8 ounces = 250 milliliters

Appendix C: 附录C

Glossary

词汇表

addend（加数）—a number to be added to another in an addition problem

addition（加法）—joining two or more numbers to get a total number

algebraic equation（代数方程）—an equation that contains variables

area（面积）—the actual surface of an object or the space it covers

attribute（特征）—a characteristic like color, size, or shape

bar graph（条形图）—a graph that uses bars to show numbers so they can be compared easily

calculate（计算）—to find an answer using mathematical processes

calculation（计算过程）—the process of finding a mathematical answer

capacity（容积）—the amount it takes to fill a container

category（范畴）—a class to which something belongs

centimeter（厘米）—a measurement in the metric system used to measure length

certain（肯定的）—something that will definitely happen

characteristic（特征）—something that identifies or sets apart an item, number, or group

classify（分类）—to organize into groups or categories

common denominator（公分母）—a number that is a common multiple of the denominators; a number that divides equally into the denominators

common factor（公因数）—a whole number that divides two or more numbers exactly

commutative property（交换律）—the sum remains the same when the order of the addends is changed; $a + b = b + a$

compare（比较）—to examine items to find similarities

compensation method（补差法）—a method of addition that teaches to add more than required and then subtract to get the correct answer

composite number（复合数）—a number with more than two factors; that is, numbers that can be evenly divided into the number

Glossary *(cont.)*

computations (计算) —to use a mathematical operation to find an answer

congruent (全等的) —having the same size and shape

coordinate grid (坐标格) —a grid containing two perpendicular axes (x-axis and y-axis) intersecting at a point called the origin (0,0)

coordinate pair (坐标对) —shown as pairs or numbers and/or letters (2,4); used to show position on a coordinate plane or map

cube (立方体) —a solid figure with six congruent square faces

cylinder (圆柱体) —a solid figure with two congruent circular or elliptical bases and one curved surface

data (数据) —factual information used as a basis for reasoning, discussion, and calculation

decimal (小数) —a number with one or more digits to the right of the decimal point

denominator (分母) —the quantity below the line in a fraction

difference (差) —the amount that remains after one number is subtracted from another

distributive property (分配律) —multiplying a number is the same as multiplying its addends by a number and then adding the products; 4 x 6 is same as (4 x 2) + (4 x 4)

divide (除) —sharing or grouping a number into equal parts

dividend (被除数) —the number being divided

divisible, divisibility (可除性) —a number that can be divided without a remainder; 10 is divisible by 5 and 2

division (除法) —the operation of creating equal groups

divisor (除数, 因子) —a number that will divide the dividend exactly

edge (边) —line segment where two faces of a solid figure meet

elapsed time (经过时间) —the passing of time

estimate (估算) —to make a rough calculation

equal (相等) —of the same value as another

Glossary *(cont.)*

equation (等式，方程式)—a mathematical statement where the left side of the equal sign has the same value as the right

equivalent fractions (等值分数)—two fractions having the same value or amount

estimate (估算)—to make a rough calculation

evaluate (估数)—a term used in algebra; to “work out” an algebra problem or find the value for it

even number (偶数)—a number that is divisible by two

event (结果)—the outcome of something that happens

expanded form (展开式)—a way to write numbers to show the place value of each digit

face (面)—the flat surface of a 3-dimensional shape

factor (因数)—a whole number that can be exactly divided into another number; a whole number that multiplies with another number to make a third number

formula (公式)—a mathematical rule written using symbols

fraction (分数)—a number that identifies part of a whole or part of a group

grid (网格)—a network of horizontal and vertical lines

hexagon (六边形)—a polygon with six sides and six angles

hour (小时)—a unit of time that is equal to 60 minutes; 24 hours equals 1 day

impossible (不可能的)—an outcome that will not happen

improper fraction (可约分数，假分数)—a fraction larger than a whole; the numerator is larger than the denominator

inch (英寸)—unit for measuring length; 12 inches equal one foot

integer (整数)—a number that can be positive, negative, or zero but not a fraction or a decimal

inverse operation (倒算法)—opposite or reverse operations; example: addition and subtraction

less than (小于)— the number to the left of the symbol is less than the number to the right of the symbol; <

line graph (线图)—uses lines to join points which represent the data

Glossary *(cont.)*

measure (测量)—the dimensions, capacity, or amount of something

measuring tool (测量工具)—item, like a measuring cup or spoon, used to measure

median (中值)—the middle value in an ordered set of values

minute (分钟)—a unit of time that is equal to 60 seconds; 60 minutes equals 1 hour

mode (众数)—the number that occurs most often in a set of data

more than (大于)—the number to the left of the symbol is greater than the number to the right of the symbol; >

multiple (倍数)—the product of a number and another number

multiplication (乘法)—the operation of adding a number to itself a number of times

multiply (相乘)—the mathematical operation where a number is added to itself a specific number of times

negative number (负数)—a number smaller than 0

number line (数轴)—a diagram that shows numbers as points on a line

number sentence (算式)—a sentence that includes numbers, operation symbols, and a greater than or less than symbol or an equal sign

numerator (分子)—the number in a fraction that is above the line

numerical data (数值数据)—data that is shown through the use of numbers

one-digit number (一位数)—a number between 0 and 9

operation (运算)—one of the four basic ways to solve a mathematical problem—addition, subtraction, multiplication, or division

order of operations (PEMDAS) (运算顺序)—the order in which operations should be completed

ordering (排序)—arranging by size, value, or amount

origin (原点)—the point where the *x*- and *y*-axis intersect on a coordinate grid; (0, 0)

parallel lines (平行线)—lines in a plane that never meet, that are always an equal distance from each other

Glossary *(cont.)*

pentagon (五边形)—a polygon with five sides

perimeter (周长)—the distance around an object or area

perpendicular lines (垂线)—lines that intersect to form right angles

pictograph (图表)—a graph using pictures or symbols to represent quantity

place value (位值)—the value of the place of a digit in a number

plot (标绘)—to place a point on a coordinate grid

polygon (多边形)—a plane two-dimensional shape having three or more straight sides

positive number (正数)—a number larger than zero

prime number (质数，素数)—a number that can only be divided evenly by itself and 1

prism (棱柱)—a 3-dimensional figure with two identical, parallel bases

probability (概率)—the chance that a given event or outcome will occur

probability scale (概率尺度)—a number line from 0 to 1 showing the chance that a particular outcome will occur; 0 = impossible, 1 = certain

product (乘积)—the amount or total when numbers are multiplied

proper fraction (真分数)—a fraction smaller than a whole; the numerator is smaller than the denominator

property (特征)—a common characteristic or feature

quadrant (象限)—any quarter of a plane divided by an *x*- and *y*-axis

quadrilaterals (四边形)—polygons with four sides

quotient (商)—the answer you receive after dividing one number by another

range (值域)—from the lowest value to the highest value in a set of data

rectangle (矩形，长方形)—a quadrilateral with four right angles and two sets of parallel sides that are equal in length

reflection (反射)—a mirror view

Glossary *(cont.)*

regrouping (重新分组)—rearranging the numbers in a problem; used when carrying in addition or subtraction

regular shape (规则形状)—a shape where all sides are of equal length

remainder (余数)—the amount left over when a number cannot be divided equally

rotation (旋转)—to turn an object

sequence (序列)—a list of numbers arranged by a particular rule

similar (相似的)—having the same shape but not the same size

simplify (简化)—to reduce the number in the numerator and denominator of a fraction to the smallest number possible

single-digit number (个位数)—a number between 0 and 9

square (正方形)—a quadrilateral with four right angles and four equal sides; the opposite sides are parallel

strategy (类别)—a plan for solving a problem or finding an answer

subtraction (减法)—an operation that takes one quantity away from another

sum (和)—the results of adding numbers

symbol (符号)—a sign that is used instead of words

systematic (有序地)—following the same order in solving problems or recording answers

table of data (数据表)—a list of data in a table that can de used to create a graph

three-digit number (三位数)—a number between 100 and 999

three-dimensional shape (三维图形)—a shape having three dimensions; length, width, and height

times table (时间表)—multiplication facts organized in a logical table

translation (平移)—move an item in any direction without rotating or flipping it

two-digit number (两位数)—a number between 10 and 99

two-dimensional shape (平面图形)—shape having two dimensions; height and length

Glossary *(cont.)*

unit of measurement（计量单位）—the standard amount or quantities of different systems of measurement

value（数值）—numeral representing worth or amount

variable（变量）—a letter or symbol that stands for a number; can change or vary depending on different values

Venn diagram（维恩图）—a diagram that uses overlapping circles or ovals to show similarities and differences among items

vertex, vertices（顶点）—points where surfaces meet; corners

volume（体积）—the number of cubic units it takes to fill a figure

weight（重量）—heaviness of an object

***x*-axis**（横轴）—the horizontal lines on a graph

***y*-axis**（纵轴）—the vertical lines on a graph

Answer Key
参考答案

Diagnostic Test (pages 2–6)

1. A 2. H 3. B 4. F
5. C 6. G 7. D 8. F
9. C 10. J 11. D 12. J
13. B 14. J 15. C 16. F
17. C 18. J 19. C 20. G
21. C 22. H 23. D 24. F
25. D 26. F 27. B 28. H
29. A 30. J

Place Value Grid (page 7)

Answers will vary.

Guessing and Checking (page 8)

Extension
135 stickers

Standardized Test Preparation 1 (page 9)

1. D 2. J 3. D
4. G 5. A
6. Answers will vary.

Use It! (page 10)

1. (10 x 4) + (5 x 4) = 40 + 20 = 60
2. (10 x 3) + (7 x 3) = 30 + 21 = 51
3. (10 x 5) + (4 x 5) = 50 + 20 = 70
4. (20 x 3) + (5 x 3) = 60 + 15 = 75
5. (20 x 20) + (6 x 20) = 400 + 120 = 520
6. (40 x 12) + (8 x 12) = 480 + 96 = 576
7. (200 x 20) = 4,000
8. (600 x 30) = 18,000
9. (20 x 6) + (1 x 6) = 120 + 6 = 126
10. (10 x 7) + (9 x 7) = 70 + 63 = 133
11. (20 x 30) + (5 x 30) = 600 + 150 = 750
12. (30 x 30) + (1 x 30) = 900 + 30 = 930
13. (40 x 20) + (8 x 20) = 800 + 160 = 960
14. (300 x 40) = 12,000
15. (800 x 50) = 40,000
16. (1,000 x 70) + (200 x 70) = 70,000 + 14,000 = 84,000
17. (2,000 x 60) + (400 x 60) = 120,000 + 24,000 = 144,000
18. (3,000 x 80) + (500 x 80) = 240,000 + 40,000 = 280,000

School Uniforms (pages 11–12)

Problem A
skirt = $6.00; sweater= $10.00.

Problem B
pants = $9.50; long-sleeved shirt= $5.50/pants= $9.50; sweater= $10.00.

Problem C
Answers will vary.

Sample Answer: 5 shorts ($30.00) and 5 short sleeved shirts ($17.50) cost $47.50.

3 pair of pants ($28.50) and 3 long sleeved shirts ($16.50) cost $45.00.

Extension
Answers will vary.

Standardized Test Preparation 2 (page 13)

1. A 2. J 3. D
4. J 5. B
6. Answers will vary.

Methods of Dividing (page 14)

1. a. 20 R3
   b. 100 R2
   c. 45 R6
   d. 206 R3
   e. 31 R6
   f. 52 R1

Methods of Dividing (page 14) *(cont.)*

2. a. 18 R4
   b. 32 R3
   c. 55
   d. 125 R6
   e. 157
   f. 268 R1

Guessing and Checking Problems (page 15)

Problem A
135, 137, and 139

Problem B
Pizza was selected by 39 students, hamburgers were selected by 13 students, and tacos were selected by 26 students.

Standardized Test Preparation 3 (page 16)

1. C 2. J 3. C
4. G 5. D
6. Answers will vary.

At the Fair (pages 17–18)

1. Group A: $24.25
   Group B: $22.25
   Group C: $49.50
2. Janine went on the Big Wheel and the Carousel.
   Malik went on either the Waltzer and the Carousel or the Big Wheel and the Ghost Train.

Answer Key *(cont.)*

Louise went on the Roller Coaster and the Ghost Train.

3. $1.15 left
4. No
5. It would cost $9.60. She would have $10.40 in change.
6. He either went on the Roller Coaster, Waltzer, and Big Wheel or the Waltzer, Ghost Train, and Hook-a-Duck.
7. $65.00
8. Answers will vary. Check student work.

Pony Express (pages 19–20)

Problem A

The ages of the two riders are 15 and 35.

Problem B

The letters weigh 2 ounces and 4 ounces. It would cost $8.00 to mail the 2-ounce letter. It would cost $12.00 to mail the 4-ounce letter. You could add an additional half ounce to the smaller letter.

Problem C

One rider rides 75 miles and the other rider rides 95 miles; 1,796 miles

Extension

22 riders

Standardized Test Preparation 4 (page 21)

1. C 2. G 3. C
4. F 5. D
6. Answers will vary.

In the Balance (page 22)

1. 18
2. 0.07
3. 6.9
4. −0.35
5. −28.5
6. 68.75

Decimal Cards (pages 23–24)

0.102, 0.12, 0.2, 0.210, 0.305, 0.31, 0.405, 0.45, 0.451, 0.55

***Guessing and Checking* Group Problems (page 25)**

Problem A

1 quarter, 2 dimes, 2 nickels, and 5 pennies

Problems B–C

Answers will vary. Check student work.

Standardized Test Preparation 5 (page 26)

1. A 2. H 3. A
4. F 5. B
6. Answers will vary.

Code Wheels (page 27)

1. 96; I
2. 0.07; N
3. 4.8; W
4. 9.6; H
5. 96; I
6. 1.3; C
7. 9.6; H
8. 1.5; M
9. 6.3; O
10. 0.07; N
11. 2.1; T
12. 9.6; H
13. 96; I
14. 3.9; S
15. 5.8; Y
16. 6.3; O
17. 1.2; U
18. 4.2; R
19. 1.48; B
20. 96; I
21. 4.2; R
22. 2.1; T
23. 9.6; H
24. 5.4; D
25. 76; A
26. 5.8; Y

The coded message is: In which month is your birthday?

Creating a Table (page 28)

Extension

If the students did not practice for one week, they would be able to recite 28 verses by the end of week 8.

Standardized Test Preparation 6 (page 29)

1. B 2. F 3. C
4. H 5. C
6. Answers will vary.

It's a Fact! (page 30)

1. a. 1, 2, 4, 8, 16
   b. 1, 2, 3, 5, 6, 10, 15, 30
   c. 1, 5, 11, 55
   d. 1, 2, 4, 5, 10, 20, 25, 50, 100
2. a. 1, 2, 5, 10
   b. 1, 7
   c. 1, 3, 5, 15
   d. 1, 2
3. a. 5
   b. 12
   c. 3
   d. 7

Answer Key *(cont.)*

4.

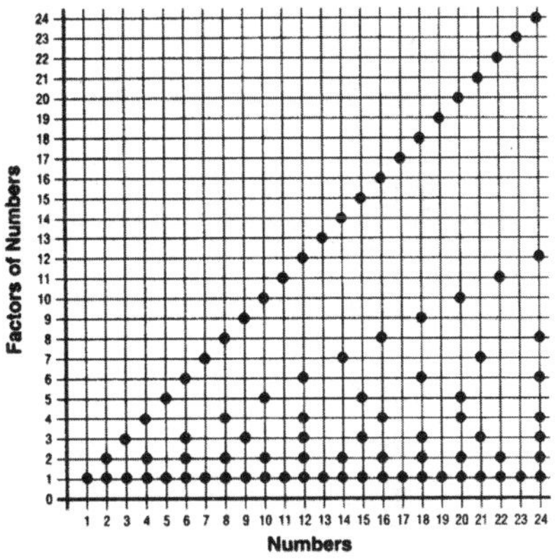

a. Answers will vary.
b. 24
c. 1, 2, 4
d. 12

Chad's Movie Rental Store (pages 31–32)

Problem A

Day Movie Is Rented	Day New Movie Is Due
Sunday	Tuesday
Monday	Wednesday
Tuesday	Thursday
Wednesday	Friday
Thursday	Saturday
Friday	Sunday
Saturday	Monday

Problem B
Rental fee: $5; no

Problem C
Rental fee: $8; Old movie–Thursday; New movie–Sunday; Yes–$4 late fee; $12 in rental fees and fines

Extension
Answers will vary.

Standardized Test Preparation 7 (page 33)

1. C 2. H 3. D
4. G 5. A
6. Answers will vary.

Hundred Chart (page 34)

Answers will vary. Check student work.

The Sieve of Eratosthenes (page 35)

1	2	3	4	5	6	7	8	9	10
11	12	13	14	15	16	17	18	19	20
21	22	23	24	25	26	27	28	29	30
31	32	33	34	35	36	37	38	39	40
41	42	43	44	45	46	47	48	49	50
51	52	53	54	55	56	57	58	59	60
61	62	63	64	65	66	67	68	69	70
71	72	73	74	75	76	77	78	79	80
81	82	83	84	85	86	87	88	89	90
91	92	93	94	95	96	97	98	99	100

The prime numbers are 2, 3, 5, 7, 11, 13, 17, 19, 23, 29, 31, 37, 41, 43, 47, 53, 59, 61, 67, 71, 73, 79, 83, 89, and 97.

***Creating a Table* Problems (page 36)**

Problem A

$t + s = 10$

$3t + 4s = 31$

$3(10 - s) + 4s = 31$

$t = 9$

$s = 1$

Therefore there are 9 triangles and 1 square.

Problem B

a. The rule is +5.
b. The rule is –1.
c. The rule is +44.
d. The rule is –11.

Standardized Test Preparation 8 (page 37)

1. D 2. F 3. B
4. H 5. C
6. Answers will vary.

Fraction Wall (page 38)

Check student work.

Adding and Subtracting Fractions (page 39)

1. $\frac{9}{10}$
2. $\frac{3}{12} = \frac{1}{4}$
3. $\frac{2}{10} = \frac{1}{5}$
4. $\frac{7}{8}$
5. $\frac{12}{13}$
6. $\frac{1}{7}$
7. $\frac{5}{6}$
8. $\frac{5}{11}$
9. $\frac{3}{3} = 1$
10. $\frac{10}{13}$
11. $\frac{4}{8} = \frac{1}{2}$ inch
12. $\frac{10}{8} = 1\frac{1}{4}$ inches
13. $3\frac{2}{8} = 3\frac{1}{4}$ inches
14. $\frac{3}{16}$ inches

Answer Key *(cont.)*

15. $\frac{8}{16} = \frac{1}{2}$ inch
16. $3\frac{8}{16} = 3\frac{1}{2}$ inches

The Popularity of Pizza (pages 40–41)

Problems A–C
Answers will vary. Check student work.

Extension
Answers will vary depending on class data.

Standardized Test Preparation 9 (page 42)

1. C 2. G 3. B
4. G 5. B
6. Answers will vary.

Simply Fractions (page 43)

1. $\frac{2}{3}$
2. $\frac{1}{8}$
3. $\frac{1}{3}$
4. $\frac{4}{5}$
5. $\frac{1}{6}$
6. $\frac{1}{12}$
7. $\frac{1}{4}$
8. $\frac{2}{9}$
9. $\frac{3}{4}$

Simply Fractions (page 43) *(cont.)*

10. $\frac{1}{5}$
11. $\frac{1}{7}$
12. $\frac{11}{41}$

Challenge
Answers will vary.

***Creating a Table* Group Problems (page 44)**

Problem A
There are 25 blue, 30 red, and 45 yellow jellybeans.

Problems B–C
Answers will vary.

Standardized Test Preparation 10 (page 45)

1. A 2. J 3. D
4. F 5. D
6. Answers will vary.

Improper Fractions (page 46)

1. $1\frac{1}{2}$
2. 2
3. $1\frac{1}{5}$
4. 25
5. $1\frac{3}{14}$
6. $1\frac{2}{5}$
7. $3\frac{1}{6}$
8. $1\frac{7}{10}$
9. $3\frac{1}{3}$
10. $7\frac{1}{2}$
11. $\frac{5}{2}$
12. $\frac{33}{8}$
13. $\frac{31}{3}$
14. $\frac{43}{6}$
15. $\frac{5}{3}$
16. $\frac{89}{8}$
17. $\frac{67}{7}$
18. $\frac{33}{10}$
19. $\frac{43}{4}$
20. $\frac{27}{5}$

Challenge
Answers will vary.

Looking for a Pattern (page 47)

Extension
Jan needs to collect 63 apples to have 36 good apples.

Standardized Test Preparation 11 (page 48)

1. D 2. H 3. B
4. H 5. C
6. Answers will vary.

On and On... (page 49)

1. 2; 5; 8; 11; 14; 17; 20; 23; 26; Rule: +3
2. 1; 3; 9; 27; 81; 243; 729; 2,187; Rule: x3
3. 2; 10; 50; 250; 1,250; 6,250; 31,250; Rule: x5
4. 27; 23; 19; 15; 11; 7; 3; –1; Rule: –4
5. 1; 2; 4; 8; 16; 32; 64; 128; 256; Rule: x2

Answer Key *(cont.)*

6. 1,024; 512; 256; 128; 64; 32; 16; 8; Rule: ÷2
7. 3; 5; 8; 12; 17; 23; 30; 38; Rule: +1 more each time
8. 80; 60; 70; 50; 60; 40; 50; 30; 40; Rule: −20, +10
9. 72; 78; 74; 70; 76; 72; 68; 74; Rule: +6, −4, −4
10. 4; 12; 36; 108; 324; 972; 2,916; 8,748; Rule: x3

Challenge

Answers will vary.

Going to the Movies (pages 50–51)

Problem A

$6.75

Problem B

$17.75; change would be $2.25

Going to the Movies (pages 50–51) *(cont.)*

Problem C

	Cost Now	Cost in 1 Year	Cost in 2 Years	Cost in 3 Years
Child Ticket	$6.50	$6.75	$7.00	$7.25
Adult Ticket	$9.50	$9.75	$10.00	$10.25
Medium Popcorn	$5.25	$5.35	$5.45	$5.55
Medium Soda	$4.00	$4.10	$4.20	$4.30

Check student work.

Extension

Answers will vary depending on class size.

Standardized Test Preparation 12 (page 52)

1. C 2. J 3. C
4. F 5. A
6. Answers will vary.

Exploring Algebraic Expressions and Equations (page 53)

1. $3x + 5 = 95$; $x = 30$
2. $5w = 40$; $w = 8$ ft.
3. $n \div 2 + 7 = 22$; $n = 30$
4. $c - 10$
5. $o + 9$
6. $p + 8$

Looking for a Pattern Problems (page 54)

Problem A

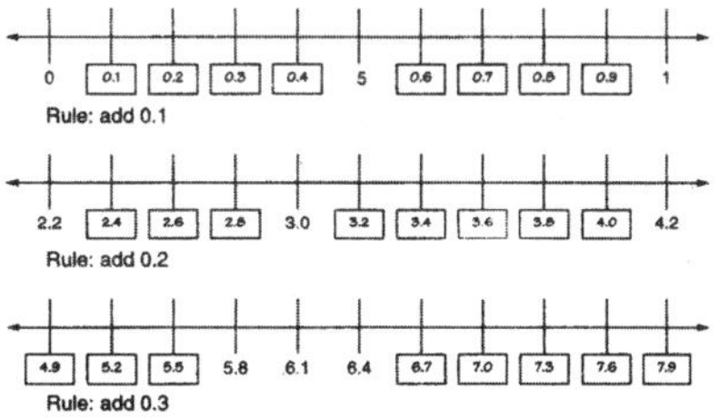

Problem B

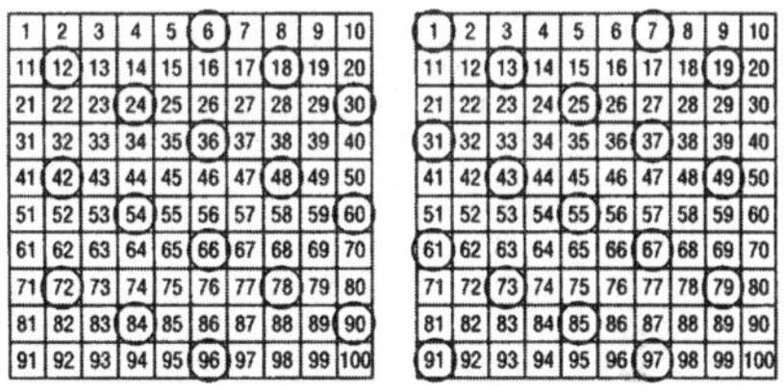

Check student work. Sample answer: The patterns are both diagonal. The only difference is the numbers that are colored.

Standardized Test Preparation 13 (page 55)

1. C 2. F 3. A
4. J 5. D
6. Answers will vary.

Variables and Equations (pages 56–57)

1. The rule is number of note cards = 8 per box or $c = 8 \bullet b$. There are 32 cards in 4 boxes. Yes, they will have enough.
2. The rule is sale price = regular price − $3.00 or $s = p - 3$. The sale price is $26.99.
3. Total miles = $2 \bullet n$

Vacation Spot	Beach	Mountains	Desert	Lake	River
Miles One Way (n)	78	267	153	241	152
Total Miles	156	534	306	482	304

4. Total students = $12 + s$

Number of Students Enrolling (s)	Total Number of Students in Class ($12 + s$)
4	16
6	18
9	21
10	22
12	24

5. The rule is sale price = regular price − $0.75 or $s = b - 0.75$. The sale price is $14.75.
6. Answers will vary.

Celebrating a Mexican Holiday (pages 58–59)

Problem A

$650; $900; a $250 difference

Problem B

$1,950; $1,350

Problem C

You can hire a 13-member band for 1 hr.; You can hire a 4-member band to play for 2 hrs.; Check student work; Range = $550; Median = $675

Extension

Check student work.

Answer Key *(cont.)*

Standardized Test Preparation 14 (page 60)

1. C 2. H 3. C
4. F 5. D
6. Answers will vary.

How Long Does It Take? (page 61)

1. a. 4 hours 40 minutes
   b. 8 hours 11 minutes
   c. 11 hours 47 minutes
   d. 10 hours 55 minutes
   e. 15 hours 48 minutes
   f. 15 hours 56 minutes
   g. 3 hours
   h. 4 hours 51 minutes
   i. 22 hours 17 minutes
   j. Answers will vary.
   k. Answers will vary.
2. 2 hours 15 minutes
3. 11:52 A.M.
4. 10:20 A.M.
5. a. 375 minutes
   b. 6 hours and 15 minutes

***Looking for a Pattern* Group Problems (page 62)**

Problem A

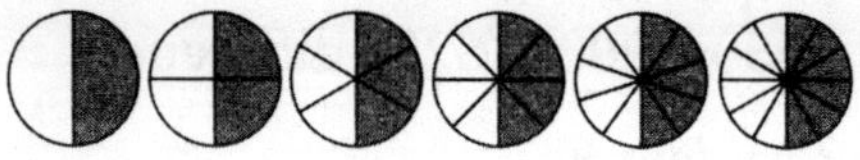

1/2 2/4 3/6 4/8 5/10 6/12

Problems B–C

Answers will vary.

Standardized Test Preparation 15 (page 63)

1. A 2. F 3. D
4. G 5. D
6. Answers will vary.

How Much? How Far? How Long? (page 64)

1. a. Dad guessed the closest.
   b. Uncle Ed was the next closest.
   c. Aunt Becky was the furthest off.
   d. 3 lb. < $5\frac{1}{2}$ lb.
      3 lb. 10 oz. > 2 lb.
      $3\frac{1}{4}$ lb. < 4 lb. 3 oz.
      $1\frac{1}{2}$ lb. < 2 lb.
      $3\frac{1}{4}$ lb. < 3 lb. 10 oz.
2. a. 1,380 yards
   b. 4,140 feet
   c. 49,680 inches
3. a. 23 feet or $7\frac{2}{3}$ yards.
   b. Yes, she has enough fabric to make 2 dresses. She needs 8 feet to make 2 dresses, and she has 27 feet of fabric.
4. a. 4 bubble baths can be made from each bottle.
   b. They would be able to finish the bottle in 1 day because there are 4 baths available per bottle.

Drawing a Diagram (page 65)

Extension

The fence would be 26 yards.

Standardized Test Preparation 16 (page 66)

1. B 2. F 3. A
4. J 5. C
6. Answers will vary.

Tiles (pages 67–68)

1. P = 10 inches
2. P = 10 inches
3. P = 8 inches
4. P = 10 inches
5. Arrangement 3 has the smallest perimeter because it has more shared edges than the other arrangements.
6. Answers will vary.

Challenge

Answers will vary.

The Summer Olympics (pages 69–70)

Problem A

The Olympic-size pool is twice as long as it is wide.

Problem B

The pool's width is 30 meters.

The Summer Olympics (pages 69–70) *(cont.)*

Problem C

Comparisons will vary. Water polo can't be played in a swimming pool because it is too long and too wide.

Extension

Answers will vary.

Standardized Test Preparation 17 (page 71)

1. D 2. H 3. A
4. J 5. C
6. Answers will vary.

Area According to Units (page 72)

1. 63 mm^2
2. 75 $in.^2$

Answer Key *(cont.)*

3. 600 yd.^{2}
4. 4 m^2
5. 2,666 cm^2
6. 420 ft.^{2}
7. 570 in.^{2}
8. 5,041 mm^2
9. 1,214.1 m^2
10. 1,617 mi.^{2}

Drawing a Diagram Problems (page 73)

Problem A

8:00–8:45	English
8:45–9:30	P.E.
9:30–10:15	Art
10:15–10:30	Break
10:30–11:15	Music
11:15–12:00	Science
12:00–12:45	Lunch
12:45–1:30	Math
1:30–2:15	Reading

Problem B

20 fence posts

Standardized Test Preparation 18 (page 74)

1. C 2. F 3. A
4. J 5. C
6. Answers will vary.

Fill It Up (page 75)

1. v = 10 cubic units
2. v = 9 cubic units
3. v = 28 cubic units
4. v = 48 cubic units
5. v = 1,443 mm^3
6. v = 238 mm^3
7. h = 3 in.
8. v = 705 mm^3
9. l = 8 ft.
10. h = 4 in.

The Tallest Mountains in the United States (pages 76–77)

Diagrams should look like this:

```
            3– – –
          1/       \\\3
        1–
      1/
   2– –
6 //////
```

Problem A

The trip will take 17 days.

Problem B

It will take 11 days to get to the top.

Problem C

6 days were spent resting.

Extension

Answers will vary.

Standardized Test Preparation 19 (page 78)

1. D 2. H 3. B
4. J 5. B
6. Answers will vary.

Name Coordinates (page 79)

Answers will vary.

***Drawing a Diagram* Group Problems (page 80)**

Problem A

There are 4 apple trees, 5 cherry trees, 2 peach trees, and 1 orange tree.

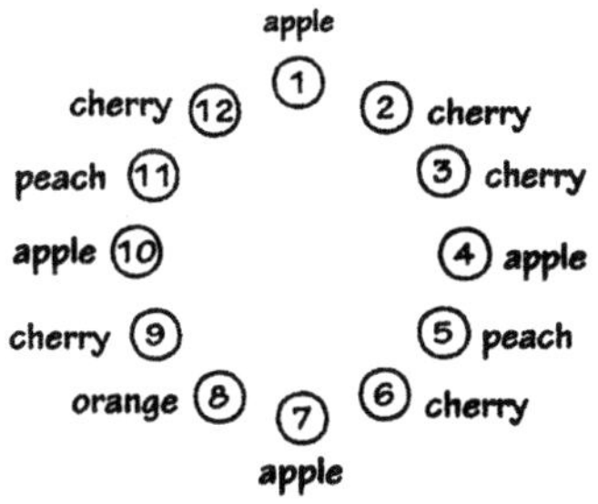

Problems B–C

Answers will vary.

Standardized Test Preparation 20 (page 81)

1. D 2. G 3. A
4. H 5. C
6. Answers will vary.

Parallel and Perpendicular (page 82)

Answers will vary.

Space to the Right (page 83)

1. a. Line (2, 3), (6, 3) is parallel to line (2, 9), (6, 9). Line (2, 3), (2, 9) is parallel to line (6, 3), (6, 9).
   b. Line (11, 1), (12, 1) is parallel to line (11, 12), (12, 12). Line (11, 1), (11, 12) is parallel to line (12, 1), (12, 12).
   c. Line (7, 0), (10, 0) is parallel to line (7, 4), (10, 4). Line (7, 0), (7, 4) is parallel to line (10, 0), (10, 4).
2. a. (5, 2)
   b. (9, 1)
   c. (2, 8)
   d. (11, 7)

Standardized Test Preparation 21 (page 85)

1. A 2. J 3. D
4. H 5. B
6. Answers will vary.

Answer Key *(cont.)*

Similar and Congruent Shapes (pages 87–88)

Section A

1–4. Check student work on page 90.

5. Congruent shapes: BCH, FEK, EDJ, GKL, HJL, BGL, BLH, EKL, and EJL
   Similar shapes: ABL, CBL, ELD, FEL, CLD, ALF, ACL, and FDL
6. Shapes BHJL, EKGL, and ELHJ are congruent. No shapes are similar.
7. Shapes AGKF, BHJE, and BGKE are congruent. No shapes are similar.
8. Shape BHLG is congruent. Shape ACDF is similar.
9. AF || KG || BE || HJ || CD; BG || FC || EJ; KE || AD || BH; AC || FD
10. AF $\perp$ FD; AF $\perp$ AC; CD $\perp$ AC; CD $\perp$ FD; BG $\perp$ HB; BG $\perp$ AD; HB $\perp$ FC; EJ $\perp$ EK; EJ $\perp$ AD; KE $\perp$ FC; AD $\perp$ FC

11–14. Drawings will vary.

Section B

Check student work on page 90.

15. Shapes BLKGA, ELHJD, and ELGKF are congruent.
16. Shapes CHJED, FKGBA, and DJHBC are congruent.

Liquids (pages 89–90)

Problem A

You can use 16 cups to empty the gallon.

Problem B

There will be 4 cups left.

Problem C

There is 1 pint remaining.

Extension

Answers will vary. Check student work.

Standardized Test Preparation 22 (page 91)

1. C 2. G 3. C
4. F 5. A
6. Answers will vary.

***Acting it Out or Using Concrete Materials* Problems (page 92)**

Problem A

Answers will vary.

Sample answer

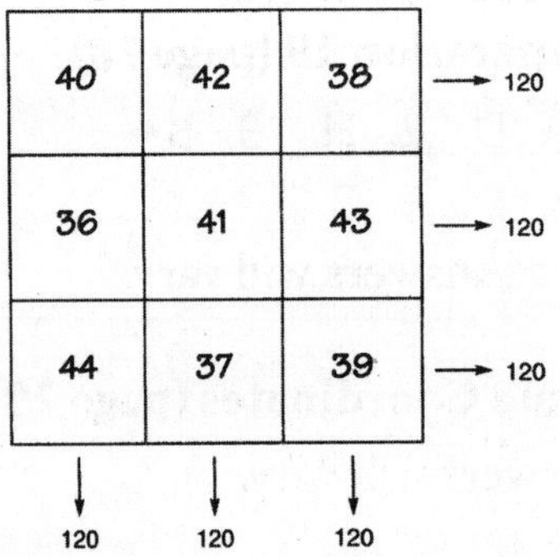

Problem B

Ethan owes his dad $1.00.

Shape Coordinates Recording Sheet (pages 95–96)

Answers will vary.

Standardized Test Preparation 23 (page 97)

1. B 2. F 3. D
4. F 5. A
6. Answers will vary.

Venn Diagram (page 98)

1. Answers will vary. Check student work.
2. Venn diagram of two-dimensional shapes:
   is not symmetrical: triangle, trapezoid
   has at least two sides of equal length: square, octagon
   outside of the two categories: circle
   middle: pentagon

The Box Factory (pages 99–100)

Problem A

Box A has 5 flat surfaces. Box B has 6 flat surfaces.

Problem B

Box A has 1 square and 4 triangles. It is a square-based pyramid. Box B has 6 squares. It is a cube.

Problem C

Box A has equilateral triangles. The base of Box A is a square because the 4 triangles have equal lengths. The area of a side of Box B is 1.27 in.^{2}

Extension

Answers will vary.

Standardized Test Preparation 24 (page 101)

1. C 2. H 3. A
4. H 5. A
6. Answers will vary.

Properties Problems (page 103)

Answers will vary. Check student work.

Answer Key *(cont.)*

3-D Shape	Number of Faces	Number of Vertices	Faces + Vertices	Number of Edges
1. cube	6	8	14	12
2. rectangular prism	6	8	14	12
3. tetrahedron	4	4	8	6
4. square-based pyramid	5	5	10	8
5. cone	1	1	2	0
6. triangular prism	5	6	11	9
7. hexagonal prism	8	12	20	18

Challenge

1. The number of faces + vertices is always two greater than the number of edges.
2. No, it doesn't work for spheres.

***Acting It Out or Using Concrete Materials* Group Problems (page 104)**

Problem A

1.

Container	1	2	3	4	5	6	7	8	9	10
No. of Blocks	2	4	3	5	3	5	2	5	3	5

2. There are 37 blocks total.
3. Containers 4, 6, 8, and 10 have 5 blocks each.

***Acting It Out or Using Concrete Materials* Group Problems (page 104)** *(cont.)*

4. Containers 1 and 7 have 2 blocks each.

Problems B–C

Answers will vary.

Standardized Test Preparation 25 (page 105)

1. D 2. J 3. C
4. F 5. A
6. Answers will vary.

Investigate It (page 106)

Answers will vary.

Standardized Test Preparation 26 (page 108)

1. B 2. H 3. C
4. H 5. B
6. Answers will vary.

Probability of Shapes (page 111)

Answers will vary. Check student work.

What Does an Architect Do? (pages 112–113)

Problem A

House Numbers	Eights in Ones Place	Eights in Tens Place
1–9	8	
10–19	18	
20–29	28	
30–39	38	
40–49	48	
50–59	58	
60–69	68	
70–79	78	
80–89	88	80, 81, 82, 83, 84, 85, 86, 87, 89
90–99	98	
100–109	108	
110–119	118	
120–129	128	
130–139	138	
140–149	148	
150		

A total of 24 houses have the digit 8 as part of the house address.

Problem B

38 houses have the digit 8 as part of their house address.

Problem C

33 houses have the digit 3 as part of their house address.

70 houses have the digit 1 as part of their house address.

Extension

Counting 1s there are 70 houses. Counting 8s there are 24 houses. When counting 1s, the hundreds place needs to be included.

Standardized Test Preparation 27 (page 114)

1. A 2. H 3. B
4. J 5. A
6. Answers will vary.

Marble Graph (pages 115–116)

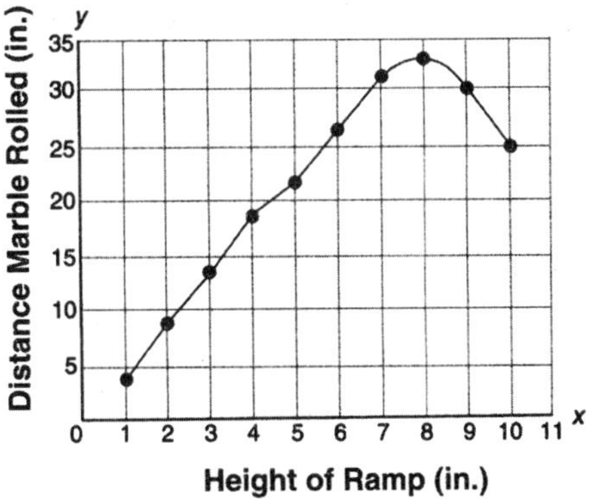

Section A

1. The distance the marble rolled reached its peak when the ramp was around 8 inches high.

Answer Key *(cont.)*

2. Answers will vary. Sample answer: It is harder for the marble to go up a steep ramp.
3. Answers will vary. Sample answer: If the x-value is 11, then the y-value is around 20 inches.
4. Answers will vary. Sample answer: If the height of the ramp was $4\frac{1}{2}$ inches, the marble probably rolled around $19\frac{1}{2}$ inches.

Marble Graph (pages 115–116) *(cont.)*

Section B

1. B 2. H 3. F
4. M 5. I 6. K
7. A 8. N 9. O
10. Q 11. C 12. E
13. L 14. P 15. D
16. G

Using Simpler Numbers Problems (page 117)

Problem A

A total of 54 cubes is needed.

Problem B

1. $4^5 = 1,024$ is larger than $4 \times 5 = 20$.
2. $3^4 = 81$ is larger than $3 \times 4 = 12$.
3. $2^6 = 64$ is larger than $2 \times 6 = 12$.
4. $5^2 = 25$ is larger than $5 \times 2 = 10$.
5. $6^3 = 216$ is larger than $6 \times 3 = 18$.

Standardized Test Preparation 28 (page 118)

1. C 2. J 3. A
4. J 5. C
6. Answers will vary.

Endangered Reptiles (pages 120–121)

Check student work.

1. There are more endangered reptiles now.
2. 126 more reptiles
3. 1,386 more amphibians than reptiles
4. 2,152 amphibians and reptiles
5. 1,287 mammals and amphibians
6. 3,324 mammals, reptiles, and amphibians
7. Questions will vary.

The Floors Under Your Feet (pages 122–123)

Problem A

You need 28 more tiles;
Total cost = $14.84

Problem B

The area of the kitchen's floor is 300 sq. ft.; The perimeter is 70 ft., so you need to buy 70 ft. of floor board.

Problem C

90 tiles are needed to cover a 180 sq. ft. area.

Extension

Answers will vary.

Standardized Test Preparation 29 (page 124)

1. C 2. J 3. C
4. J 5. B
6. Answers will vary.

Flipping Coins (pages 125–126)

1–2. Check student work.

3. $\frac{1}{4}$
4. $\frac{1}{4}$
5. $\frac{1}{2}$

6–9. Check student work.

Will It or Won't It? (pages 127–128)

1. $\frac{1}{2}$
2. a. $\frac{6}{12} = \frac{1}{2}$
   b. $\frac{4}{12} = \frac{1}{3}$
   c. $\frac{2}{12} = \frac{1}{6}$
3. a. $\frac{5}{10} = \frac{1}{2}$
   b. $\frac{4}{10} = \frac{2}{5}$
   c. $\frac{9}{10}$
   d. $\frac{2}{10} = \frac{1}{5}$
   e. $\frac{8}{10} = \frac{4}{5}$
4. $\frac{1}{2}$

Will It or Won't It? (pages 127–128) *(cont.)*

5. a. $\frac{1}{6}$
   b. $\frac{3}{6} = \frac{1}{2}$
   c. $\frac{3}{6} = \frac{1}{2}$
6. a. Answers will vary.
   b. Answers will vary.

Answer Key *(cont.)*

7. a. $\frac{2}{10} = \frac{1}{5}$

   b. $\frac{2}{10} = \frac{1}{5}$

   c. $\frac{4}{10} = \frac{2}{5}$

   d. $\frac{1}{10}$

   e. $\frac{1}{10}$

8. a. $\frac{1}{11}$

   b. $\frac{4}{11}$

   c. $\frac{4}{11}$

   d. $\frac{2}{11}$

Using Simpler Numbers Group Problems (page 129)

Problem A

There are 15 footprints after 3 jumps. There are 50 footprints after 10 jumps. There are 100 footprints after 20 jumps.

Problems B–C

Answers will vary.

Standardized Test Preparation 30 (page 130)

1. D   2. G   3. C
4. G   5. C
6. Answers will vary.

Match It! Game Cards

1. 87,493
2. 31 R 8
3. $5.90
4. 0.67
5. 600,000 or 6 hundred thousands
6. 4.3
7. 69
8. 66 R5
9. $34.24
10. 0.06
11. $6.60
12. 153
13. 37, 42, 47
14. 63 $in.^2$
15. $\frac{4}{5}$
16. 24 units squared or 24 $units^2$
17. 275
18. 4 pounds

Find It! Game Cards

1. 18, 15
2. 6
3. $7.15
4. 96 $in.^2$
5. 8,069,090
6. 52 stamps
7. (1, 3)
8. $\overleftrightarrow{ER}$ and $\overleftrightarrow{JK}$
9. isosceles
10. 75
11. 84 pizzas
12. 96 cm^2
13. (6, 2)
14. triangle
15. 28 cm
16. 0.6
17. 180 feet cubed or 180 $ft.^3$
18. Alaska

Solve It! Game Cards

1. (5, 2)
2. 144 ft.
3. 80
4. $\overleftrightarrow{CD}$
5. 9 units cubed or 9 u^3
6. 90 in.
7. 0.95
8. $\frac{3}{8}$
9. 80°
10. 17
11. $\overline{PL}$
12. 0.181
13. $\frac{4}{9}$
14. 14
15. 1, 2, 3, 4, 6, 8, 12, 24
16. 21
17. 36 $in.^2$
18. 10°

Notes